Digital Dead End

Digital Dead End

Fighting for Social Justice in the Information Age

Virginia Eubanks

The MIT Press
Cambridge, Massachusetts
London, England

For information about special quantity discounts, please email special_sales@ mitpress.mit.edu.

This book was set in Stone Sans and Stone Serif by Toppan Best-set Premedia Limited. Printed and bound in the United States of America.

Library of Congress Cataloging-in-Publication Data

Eubanks, Virginia, 1972–
Digital Dead End : Fighting for Social Justice in the Information Age / Virginia Eubanks.
   p.   cm.
Includes bibliographical references and index.
ISBN 978-0-262-01498-4 (hardcover : alk. paper)
1. Technology—Sociological aspects. 2. Technology—Sex differences.
3. Technology and women. I. Title.
HM846.E93   2011
303.48′34082—dc22
                                                             2010018810

10 9 8 7 6 5 4 3 2 1

To the women of the YWCA of Troy-Cohoes community.

Without your insight, generosity, and humor, this book would not exist.

Contents

Author's Note

The author will donate half of the royalties from the sale of this book to the Popular Technology Workshops, a grassroots organization based in Troy, New York. The Popular Technology Workshops are dedicated to fostering a vibrant people's movement to promote economic equity, political participation, and social justice in the information age. More information about the organization can be found at <http://www.digitaldeadend.com>.

Acknowledgments

This book was inspired and enriched by the work of countless people, and I am deeply indebted to the rich and vibrant community of research and activism that grew up around it. It goes without saying that this work would be inconceivable without the remarkable generosity and resourcefulness of the women of the YWCA of Troy-Cohoes community. I would especially like to thank my research collaborators in WYMSM, including Nancy D. Campbell, Coffee, Jessica Constantine, Cuemi Gibson, Ruth Delgado Guzman, Cosandra Jennings, Chitsunge Mapondera, Patty Marshall, Christine Nealon, Zianaveva Raitano, Jennifer Rose, and Julia Soto Lebentritt. I hope I have represented our collective work faithfully. Special thanks to Isaac and Mark, who kept us smiling through long afternoon meetings.

Thanks to Pat Dinkelaker, an early and steadfast champion of this project, and executive director of the YWCA for its duration. Her leadership, vision, and kindness enriched the lives of thousands of women at the YWCA and started me on a new path. Thanks to the wonderful staff of the YWCA during my time there: Sue Alberino, Bonnie Benson, Jes Constantine, Ronnie Cush, Liz Girolami, Sue Kemnitzer, Kelly Klein, Christine Nealon, Joni Richter, Sarah Smiley, and Andrea Walters. My thanks also go to interviewees, participants in public programs, tech lab and Resource Center users, residents, and other members of the YWCA community for the honesty and generosity with which they shared their experience. They have been identified only by pseudonyms, so I won't blow their cover here. You know who you are.

My thanks also go to the members of Our Knowledge, Our Power: Surviving Welfare, the grassroots antipoverty and welfare rights organization that grew out of the work recounted here, for providing continued insight and inspiration and for keeping me accountable. Members include Linda Adams, Aalim Ammar, Anne Bink, Devona Brown, Hee Jeong

"Robyn" Chung, Jes Constantine, Roberta Farrell, Mishel Filisha, Deborah Ford, Cuemi Gibson, Catherine Gordon, Kenneth and Kenny Harris, Kizzy Howell, Molly Hussey, Victoria Kereszi, Penny Lane, Nadya Lawson, Tiffany Lherisson, Andrew Lynn, Patty Marshall, Shihoko Nakagawa, Christine Nealon, Ibi Oduyemi, Mary Pimble, Denise Roe, Glen Taylor, Donna Tenney, Shaneeka Thrasher, Maggie Torres, Kristin Turano, Jennifer Vasquez, Dametriss Walker, and Kim Wheeler.

I owe a great debt to my editor at MIT Press, Marguerite Avery, who challenged me to produce a book that would appeal to both academic and popular audiences, providing me space to rediscover the joy of writing for a broad readership. Thanks to four anonymous reviewers, whose input greatly improved the book. I am grateful to the members of Rebecca Wolff's Hybrid Writing Workshop at the New York State Writers' Institute for giving me a much needed push early in my rewriting. Thanks are also due to various incarnations of the Ladies Writing League—Anne Bink, Jennifer Burrell, Amy Halloran, Alethia Jones, Patricia Strach, and Meisha Rosenberg—who fed me, supported me, and provided critique and encouragement throughout the process. Amy, Alethia, and Meisha read every word of the manuscript, sometimes several times, as I struggled to put the whole story together in the final year of writing, and I owe them more than I can say.

I am extremely grateful to the staff, interns, and visitors at the Highlander Research and Education Center in New Market, Tennessee, especially Kristi Coleman, Mónica Hernández, Pam McMichael, Jenny Paris, Rob Reining, Sheena Sood, Elandria Williams, and Susan Williams, who provided me sanctuary at two important times in this process. Thanks to Sanford Schram, who rode in like a knight in shining armor at a particularly tense moment, and to Joe Soss, one of my most enthusiastic and consistent supporters. Thanks to *Times Union* business writer Larry Rulison and feminist economists Drucilla Barker and Suzanne Bergeron for providing timely and provocative comments on chapter 4.

This project spanned my time at two unique academic institutions and extended over a period of intense change in my personal and professional life. I would not have made it through without my colleagues and friends at Rensselaer Polytechnic Institute and the University at Albany, State University of New York. At Rensselaer, I owe a debt of gratitude to my dissertation committee, who helped me start this project: Rayvon Fouché, Davydd Greenwood, David Hess, and Langdon Winner. I thank my lucky stars that Nancy D. Campbell, my dissertation chair, decided to take a chance on a "high-risk, high-yield" doctoral thesis, and stuck with me through all the ramifications of that decision. I am proud and honored to call you a friend and mentor.

The Community Outreach Partnership Center (COPC) at Rensselaer—Frances Bronet, Nancy D. Campbell, Ron Eglash, Branda Miller, Barb Nelson, and Steve Pierce—provided funding and encouragement through the research stages of this project. I thank my colleagues in Science and Technology Studies at Rensselaer: Michael G. Bennett, Jean-Francois Blanchette, Camar Diaz-Torres, Nicole Farkas, Jill Fisher, Jenrose Fitzgerald, Kim and Michael Fortun, Art Fricke, Abby Kinchy, Torin Monahan, Dean Nieusma, Jason Patton, Hector Postigo, Erich Schienke, Jeanette Simmonds, and Alex Sokoloff. Your friendship and insight helped me figure out who I am as a scholar and an activist.

I am thrilled that I found a home in the Department of Women's Studies at University at Albany, SUNY. I am inspired and encouraged by a wonderful group of colleagues in the department and across the campus: Chris Bose, Maia Boswell-Penc, Janell Hobson, Alethia Jones, Fernando Leiva, Vivien W. Ng, Gwen Moore, Marjorie Pryse, Loretta Pyles, Bonnie Spanier, Glenna Spitze, Patricia Strach, and Barbara Sutton. Special thanks go to Julie Novkov, my mentor and professional idol, for shepherding me through difficult times and evening my keel. Though there are too many to list here, I draw endless motivation and energy from all my wonderful students in the Women's Studies department. I am privileged to watch you become sharper thinkers, better advocates, and more fearless fighters for justice.

It is easier to be courageous in concert, so I draw great strength from the community of feminist activists in the Capital Region. Those I have not already mentioned include Jaclyn DeChiro, Q Diamond, Jackie Hayes, Nadya Lawson, Naomi Jaffe, Cricket Keating, Roxanne Ramlall, Carmen Rau, Barbara Smith, and Fiona Thompson. I owe my wonderful friends Nina Baldwin, Tanya Bastone, Celia DuBose, Adam Frelin, Jackie Goss, Amy Halloran, Rachel Havrelock, Niki Haynes, Alethia Jones, Penny Lane, Steve Rein, and Kate Sipher many phone calls, and thank them for helping me keep body and soul together. Thanks to Nick Matulis for putting a roof over my head, and for always asking how the writing's going. Thanks to my family—Carole Eubanks, Hardy Eubanks, David Deigan, Tommy and Brenda Eubanks, and Daphne, Patrick, and Elizabeth Haugh—for their support of this project and their endless patience with me.

Finally, I am especially and eternally grateful to Jason Martin, my first reader and closest friend, who teaches me daily how to live a life of courage, integrity, and honesty.

This research was funded in part by a HUD Community Outreach Partnership (COPC) grant and a National Science Foundation dissertation improvement grant (NSF 0322525). The manuscript was completed with the help of a Dr. Nuala McGann Drescher Leave Award from the Joint Labor-Management Committees of New York State and United University Professions. Any opinions, findings, and conclusions or recommendations expressed in this material are those of the author and do not necessarily reflect the views of the Department of Housing and Urban Development, the National Science Foundation, the Joint Labor Management Committees, or United University Professions.

Introduction

In his first address to Congress, in February 2009, President Barack H. Obama called on the redemptive power of science and technology to help pull the nation out of its deepest economic crisis since the Great Depression. Even before his inauguration, in a January 15 summary of the American Recovery and Reinvestment Act (ARRA), the Committee on Appropriations argued forcefully, "We need to put scientists to work looking for the next great discovery, creating jobs in cutting-edge technologies, and making smart investments that will help businesses in every community succeed in the global economy" (Obey and Committee on Appropriations 2009). ARRA set aside $16 billion, about 3 percent of proposed investment spending, for science facilities, research, and instrumentation, and to expand broadband access in rural and underserved areas. In his address, President Obama explained, "We have . . . made the largest investment in basic research funding in American history—an investment that will spur not only new discoveries in energy, but breakthroughs in medicine, science and technology" (Obama 2009). This commitment, he argued, is crucial to ensure American competitiveness and leadership on a global scale.

President Obama's proposed investments in American education, infrastructure, sustainable energy, and health care are certainly well placed and long overdue. But I argue in this book that continued emphasis on the development of science and technology as *the* route to greater prosperity and equality for all Americans is a familiar but dangerously underexamined species of magical thinking. In psychology, magical thinking is the belief that merely thinking about an event in the external world can cause it to occur, a delusion often present in very young children and schizophrenics. Many of us in the United States have engaged in a massive, collective, consensual hallucination about the power of technology, particularly information technology (IT), to "level the playing field," create broad-based economic and social equality, and nurture transparency and accountability in democratic governance.

This magical thinking has its root in an incomplete picture of the impacts of IT and technology-driven economic development schemes in our communities, a myopia shaped by race, class, and gender inequality. This shortsightedness in turn skews our policy responses to issues of high-tech equity and, in many cases, creates policies and institutions that deepen inequality rather than alleviate it. We need to expand and clarify our vision of equity in an information age. Massive investment in science and technology without simultaneous investment in a more just society is an investment in *increasing* political and economic inequality. If robust democracy, broadly shared prosperity, human rights, and equity are important to us as a society, we must reject our magical thinking and look with clear and courageous eyes at the real world of IT, our shared technological present.

In retrospect, the ongoing global financial crisis makes the "new economy" platitudes of the last decade—the high-tech economy is a rising tide that will lift all boats, the information age ushers in a new era in which knowledge triumphs over matter, globalization equalizes the playing field and flattens hierarchies around the world—seem delusional. But magical thinking is deeply infectious. Throughout the process of researching and writing this book, I found it difficult to separate myth from reality myself. A collection of newspaper clippings in my research files illustrates the schizophrenia that swept my home, the Capital Region of upstate New York, in the time between the peak of the dot-com bubble (2000) and the beginning of the worldwide economic downturn (2008). The headlines make for some interesting pairings. "Tech Valley Firms Poised for Growth" is right before "Gap between Rich and Poor Grows Statewide." "High-Tech Hopes Increase Optimism" is close to "Nearly 800,000 People Will Lose Benefits after Christmas." "Education Engine Helps Power the Capital Region" is chastened by "Colleges Cut Budgets as Market Erodes Endowments." The most compelling article sums it all up in a single 2003 headline, "Optimism on Rise in Tech Valley, Hiring Remains Flat."

This is a book about the myths and realities of the high-tech global economy for people who live in persistent poverty in the United States. More specifically, it is about how magical thinking led to increased economic, social, and political inequalities during a regional effort to create Tech Valley here in the Capital Region of New York. As a corrective against magical thinking, I have labored to keep myself grounded in national, state, and local economic reality and in the day-to-day experiences of women and families struggling to meet their basic needs in my hometown of Troy. The story I tell is multifaceted, partially because it responds to the dualisms of the information age. But it is also multivocal because the kind of research I engaged in—a collaborative, long-term approach called participatory action research—included the input and analysis of dozens of people from a wide variety of backgrounds, cultures, and experiences. One way to guard against magical thinking, to get out of the habit of denial, is to compare your perceptions with those of others around you, especially those who do not share your social experience of the world. I used participatory methods so that it would be harder to maintain any kind of convenient fiction, testing theories and policy models collectively through practical community action.

Most of the collaborative research that supports this book took place within a grassroots organizing group at the YWCA of Troy-Cohoes, an organization that is both home to ninety women seeking to craft the lives they want for themselves and part of a national membership movement to empower women and eliminate racism by any means necessary. The group was called WYMSM (Women at the YWCA Making Social Movement—we pronounced it *wim-sim*), and was made up of myself,

Figure 0.1
The first incarnation of Women at the YWCA Making Social Movement. From left to right: Nancy D. Campbell, Marva Ray, Jenn Rose, Coffee, Patty Marshall, Jes Constantine, Virginia Eubanks, and Chitsunge (Chris) Mapondera.
Photo: Christine Nealon

Nancy D. Campbell, "Coffee" (a pseudonym), Jes Constantine, Cuemi Gibson, Ruth Delgado Guzman, Cosandra Jennings, Julia Soto Lebentritt, Chitsunge (Chris) Mapondera, Patty Marshall, Zianaveva Raitano, and Jennifer Rose. Our mission statement read:

As an initiative of the YWCA of Troy-Cohoes community, *Women at the YWCA Making Social Movement* (WYMSM) seeks to use technology as a tool of social change. This community-building collaboration creates projects that help women build awareness of existing resources, knowledge, and experience; precipitate resource sharing and development; and provide supportive encouragement to learn from others' experiences through technological tools and social network building.

WYMSM members' voices—drawn from interviews, public events, meeting recordings, and written correspondence—animate this book and shaped my interpretation of additional data collected through public workshops, classes, and events, and in twenty-nine interviews with YWCA residents, staff, community members, and other Troy residents. There are many voices fighting to be heard in this account, and many different interpretations that vie for attention. I have attempted to be true both to my own analysis and interpretation as an author and to the voices of the incredible community of women I worked with throughout the project. I believe the result portrays the realities of living on the ground, with both eyes open, in the information age.

A Roadmap

Digital Dead End began with my attempts to create technology training programs with women living in the YWCA community, efforts that were interrupted by their counterattempts to articulate ambivalence about technology, describe their everyday interactions with it, and express their hopes for a more just future. Their stories challenged my preconceptions, overturned the central tenets of digital divide policy, and shattered the familiar illusion that low-income people are somehow information or technology poor. Their insights forced me to reach beyond the most common model of high-tech equity in the United States—universal access—to explore the relationship among technology, politics, citizenship, and social justice.

This roadmap is intended to help readers chart their own path through the interlocking and complicated stories that make up this book. To begin, in chapter 1, I offer four different origin stories: a bit of my own history, a moment from a workshop at the YWCA that catalyzed major changes in my thinking, a description of a collaborative project undertaken at the

YWCA, and a reflection on love and social action. These points of entry should give the reader a clearer idea of why I approached this research the way I did, what it looked like in practice, and the challenges and advantages participatory research offers.

IT policy and activism often assume middle-class values and experiences, obscuring or neglecting the unique insights and struggles of poor and working-class people.[1] In chapter 2, I discuss the conceptual or theoretical stumbling blocks that keep scholars from understanding the relationship between technology and poverty more fully, including limited ideas about equity and justice, lack of attention to social location, narrow conceptions of citizenship, overly static definitions of technology, and inadequate methodology. In chapter 3, I discuss how these conceptual stumbling blocks have led to public policy that neither takes poor and working-class people's real-life experiences into account nor adequately provides for social justice in the information age. These oversights and omissions are particularly evident in policies directed toward bridging a presumptive digital divide, which, I argue, are trapped in a distributive paradigm that sees all high-tech equity issues as distributive issues.

As a corrective to the oversights described in chapters 2 and 3, I offer the stories and analysis of women living in the YWCA community. In chapter 4, I discuss their experiences of the information economy as our hometown of Troy, New York, seeks a place in a regional economic development initiative called Tech Valley. I also provide evidence that low-income women in Troy participate in the information economy in huge numbers, and I describe their experiences in both low-wage, high-tech jobs and in the service and caregiving industries that make high-tech growth possible. In chapter 5, I explore the interaction of women in the YWCA community with IT in the social service office and investigate the political lessons they learn when dealing with technologies of state administration. The experiences of women in the YWCA community with the information economy and technologies of state administration directly contradict the widespread belief that poor and working-class women lack access to technology. In fact, they describe their lives as characterized by technological ubiquity—technology shapes their workplaces, community institutions, and political experiences. But, unlike many of their middle-class counterparts, their encounters with IT and the high-tech economy tend to be exploitative and limiting, increasing their economic vulnerability and political marginalization.

In this context, the ambivalence women in the YWCA community sometimes expressed toward technology is both understandable and reasonable. But this ambivalence is double-edged: as more economic, political, and cultural power is routed through IT networks, it is increasingly important to create technologies that integrate broad-based democratic participation and decision making. The second half of the book explores an approach to creating a broadly inclusive and empowering "technology for people," an approach I call *popular technology*. Popular technology assumes that all people have a rich array of experiences with technology, shaped by their social location, and that these experiences provide a valuable resource for thinking collectively and critically about the relationship among technology, politics, citizenship, and social justice. Popular technology entails shifting from vocational approaches that teach technological skill to popular education approaches that focus on nurturing critical technological citizenship. This shift can have a significant impact on scholarly, policymaking, and social justice work, as well as improve the everyday lives of poor and working-class women and their families.

In chapter 6, I discuss the three models on which popular technology is based—popular education, participatory action research, and participatory design—and describe three popular technology projects undertaken at the YWCA between 2001 and 2004. In chapter 7, I explore the lessons learned in these projects and other WYMSM activities to develop a theory of cognitive justice for the information age. The concepts of cognitive justice and critical technological citizenship offer us a way out of the trap of the distributive paradigm because they broaden the scope of justice beyond access or distribution to encompass freedom from oppression of all kinds. In the conclusion, I synthesize the lessons of the rest of the book to offer a programmatic take on popular technology and to develop a high-tech equity agenda that serves *all* people.

In addition to the central narrative of the book, I have included a wealth of extra information to give readers a sense of how participatory research works, and to encourage readers to give popular technology a try in their own communities. Appendix A is a thorough account of our research methodology, which clearly lays out how the research project was designed, how data were collected and analyzed, and how my collaborators' insights and analyses were integrated into the final book. Appendix B focuses more narrowly on WYMSM's meetings, providing agendas and supporting materials to give the reader a rich sense of what the group did on a week-to-week basis. In addition, the reader is introduced to WYMSM throughout the book in member profiles: brief, mostly first-person narratives that WYMSM members and I co-wrote in 2009. Appendix C provides agendas and supporting information for three popular technology workshops offered in different contexts between 2001 and 2009. This information is included in the hope that readers will be sufficiently intrigued to offer the workshops—focused on the self-sufficiency wage, the social service system, and the relationship between technology and social justice activism—in their own communities. Finally, appendix D briefly lists all the popular technology projects we undertook at the YWCA of Troy-Cohoes in order to give readers a sense of activities that did not make it into the book because of space limitations.

I have provided this information because I hope the book will reach a wide audience of scholars, activists, and policymakers. Not all audiences will be equally interested in all chapters. Primarily academic audiences might be more interested in the theoretical work detailed in chapter 2, for example, while activist readers might go directly to the second half of the book and the appendices. I hope that my efforts to tell all these stories in the same book will stretch ideas of what a scholarly book is capable of achieving and provide space for knowledge and analyses that have been truly co-created by engaged academics, activists, community-building organizations, neighbors, and friends. I have tried to weave many threads into a single fabric, because what we lack in high-tech equity work is a holistic picture of the relationship between technology and inequality, one not shaped by the magical belief that simply wishing for justice makes it so.

1 Four Beginnings

I was born in Dallas, Texas, and brought up in Ho-Ho-Kus, New Jersey, an almost entirely white, middle-class suburb of New York City. Growing up, my experience of the wider world was pretty limited, but my mother, despite her attempts to downplay her working-class East Texas roots in the context of my dad's more patrician banking family, had a strong sense of social justice, equity, and fairness. One of her favorite stories about me concerns an incident that occurred when she was driving me to preschool in Austin, Texas. It was 1976. I was four, and she had recently become a subscriber to *Ms. Magazine*. On that day, I asked about a big building we were passing, one with huge columns like a temple. She told me that it was a Masonic Hall, slyly adding that Masons didn't allow women into their group. When she asked me what I thought about that, she swears that I yelled "That's not fair!" and demanded she pull the car over so that I could go in and talk some sense into them.

Despite my natural inclination toward speaking out, I was raised in a culture of silence. Middle class, white, suburban, and deeply affected by a family member's alcoholism, I always felt as though a secret lay simmering just below the surface of our outwardly calm and prosperous life. I have since found out that this is a pretty common experience for middle-class white people who become antiracist and antipoverty activists later in life. Many of us describe growing up as worried or angry kids, struggling against the shoddy logic and emotional repression that sustain illegitimate power relationships and underwrite white supremacy and economic exploitation.[1] My parents are deeply decent people; they vocally challenged discrimination, worked on political campaigns, and raised two strong-willed, independent daughters. But our family was caught in the web of color-blind racism and class-blind classism: while my parents would not have tolerated a racist or classist joke, they had no close friends of color, and our family never discussed the source—or the impacts—of our money or privilege.

Growing up in a bubble of privilege made me intensely curious; explanations for the structure of our society never quite rang true. Like a little detective, I cross-checked facts, grilled witnesses, followed hunches, researched and read assiduously. I scratched surfaces, was slow to accept the party line, questioned everything. A relentless kid, I realized early that I was living in a sea of what feminist sociologists of knowledge call sanctioned ignorance, a set of culturally endorsed falsehoods and half-truths we are asked to swallow to maintain the status quo. A defiant kid, I was not about to let ignorance and lies define my life. I practiced dissent from a very early age—when I was barely ten years old I wrote a two-page newsletter arguing why people shouldn't hunt deer, duplicated it, and put it on the windshields of every car in my elementary school parking lot. I also showed early political acumen: I managed to convince the elementary school to let me use the photocopier in the front office to reproduce it!

I was bookish, so I did a lot of reading—the *Communist Manifesto*, *The Autobiography of Malcolm X*, and John Stuart Mill's *On Liberty* were in my adolescent library, alongside *Are You There God? It's Me, Margaret*; *Black Beauty*; and *A Wrinkle in Time*. In seventh grade, after my parents' divorce had somewhat narrowed our economic circumstances, I mentioned Karl Marx in a discussion of inequality in an English class, only to have my teacher tell me that he was the "guy who made up the [Aryan] Superman" and that he was, more or less, a Nazi. At the time, I did not know that my teacher was mixing up Marx and Nietzsche and getting Nietzsche wrong, to boot, but I did know, unequivocally, that she was wrong. Two things became clear. The first was that people in my suburban hometown didn't want to talk about economic inequality *at all*. The second was that the things I was being taught in the classroom didn't reflect my experience of the world or my innate sense of right and wrong.

That radicalizing moment also earned me the nickname "Pinko" among my peers, which stuck with me for five long years. After getting the nickname, I followed a pretty predictable path for a middle-class white girl with a budding political consciousness: animal rights activism and vegetarianism in my early teens, environmental activism from age fifteen to seventeen, a quick break for black turtlenecks and moody boyfriends, and then college at a liberal, gradeless state university in California. It was during college that I got involved with community media through a local noncommercial radio station and discovered the Internet, though it was largely an esoteric and specialized realm during my college years, which spanned the early 1990s.

During that time, my activism and thinking about justice began to shift and deepen. I discovered feminism when a women's studies class gave me language to articulate long-held beliefs about sex and gender inequalities. I engaged in antiwar and anti-imperialist organizing after the first invasion of Iraq. I took my first steps into antiracist and civil rights work after hitchhiking to San Francisco to hear Angela Y. Davis speak at the Western Regional Organizing Conference Against the War in 1991; it was a profoundly life-changing experience. While my college campus was diverse in terms of race, nationality, sexual orientation, and beliefs, it was not terribly economically diverse, and though I met a few Marxists and anarchists, and sought out collectives and co-ops in town, there wasn't much of a conversation going on about economic inequality.

And then came the rumblings of the information revolution. There in the heart of the Silicon Valley, while working as the development director for a community radio station, I discovered this fascinating new thing called the World Wide Web. I hacked my way through HTML, started making Web sites (for the Mosaic browser!), and moved up the coast to San Francisco to start my post-college life in 1995. Those were strange days in the Bay Area. For a young woman like me with racial and economic privilege, a college degree, no family obligations, and some working knowledge of computers, it was a remarkable time of freedom and excitement. I set myself up as a freelance Web site developer, found a $300 per month room in the Mission District, and started one of the first cyberfeminist 'zines, a short-lived snarky online periodical called *Brillo*.

But even in the heady atmosphere of the dot-com boom, it would take a powerful brand of denial to not see that something was amiss in the middle of the Silicon Valley miracle. Though my vision was limited by my privileged social and economic position, I was not blind. It was clear to me at the time that I was part of the massive wave of gentrification that swept through San Francisco neighborhoods like the Mission, South of Market, Hayes Valley, and the Western Addition. Public housing began to disappear, replaced by coffee shops, Internet cafes, and the kind of stores that display two items of clothing in a big white room. In the three and a half years I lived in San Francisco, the vibrant diversity of the city waned visibly and rents in my neighborhood tripled.

In the mid-1990s, in the circles I was running in, it was not unusual for people to ask you at parties, only half ironically, "Have you made your first million yet?" It was, many believed, the American Dream manifest: all you needed was a good idea, some sweat equity, and a garage, and the digital economy would bestow on you its mighty gifts. I understood the itch for the million. Straightforward greed was not what was tying my brain in knots. What I had trouble wrapping my head around was Silicon Valley's unique way of combining utopian fervor with blatant dissociation from reality, a cognitive dissonance that led me to a personal crisis of conscience and eventually drove me out of the Bay Area.

People around me seemed to believe that the high-tech economy was going to lift all boats—lead to better outcomes for everyone—but they were ignoring the obvious evidence of increasing economic inequality that I saw around me every day. How could people simultaneously think they were all going to get filthy rich *and* make the world a better place for everyone? The people commending the economic miracle in Silicon Valley seemed to be suffering from a kind of collective, consensual blindness, blocking out the gentrification, the skyrocketing rents, and the toxic environmental toll of high-tech industry. The increasing disparities were evident if you only had the will to look.

The solutions I found at the time, and the contributions I thought I could make, focused on access to technology. I believed that one of the key ways to mitigate the more disastrous impacts of the high-tech economy was to make the tools of the information revolution more widely available across disparities of gender, race, class, language, ability, and nationality. I began volunteering at Plugged In, a well-known community technology center in the Whiskey Gulch neighborhood of East Palo Alto, the poorest city in San Mateo County. Whiskey Gulch was an economically challenged but culturally rich neighborhood down the street from Stanford University, a community squeezed by gentrification pressures, education system short-comings, and a lack of stable, living-wage jobs. Plugged In provided youth from the community computer access, technology classes, and employ-ment training at its University Avenue address until 1999, when developers razed East Palo Alto's downtown, including Plugged In's original home, and replaced it with a Four Seasons Hotel, a convention center, and an IKEA store.

Back in the Mission District, I started free Internet and World Wide Web literacy classes for poor and working-class women through a community arts organization called Artists' Television Access. The classes concentrated on larger social issues—the Internet's birth in the defense industry, eco-nomic justice issues in the neighborhood, and gender issues online—as well as practical skills, such as using the Internet and the Web to find information, HTML authoring, and graphic design. But I had doubts that these piecemeal efforts could address the systemic, widespread economic inequalities I was witnessing. What drove me back east and into graduate school was a combination of this concern—that my activism was not really addressing the root causes of economic disparity in the high-tech economy—and the steadily increasing feeling that I was going crazy. Why did I insist on examining the goose laying the golden eggs while everyone else was drinking lattes, doing yoga, and cashing in their stock options?

So, in 1997, I fled the triumphant arrival of the "new economy" in Silicon Valley and went to live beside the Hudson River in the historic city of Troy, New York. My experiences in the Bay Area traveled east with me and remained on my mind. These formative experiences—my work in community technology centers, the publication of *Brillo*, and my experiences with magical thinking during the Silicon Valley "miracle"—mark the beginning of this book. I was a committed community technology practitioner for nearly ten years, and I believed that access to technology was a fundamental social justice issue in American cities.

I was wrong.

In this book, I try to explain the source of my misunderstanding and describe the changes that took place in my thinking between 2001 and 2004, when I was working in the YWCA community, engaged in the research that became the basis for this book. In 2001, as part of my doctoral work in science and technology studies, I set out to construct community-based technology training programs with an emphasis on peer education and the design of locally relevant tools. This work took place at the Sally Catlin Resource Center and later its associated technology lab, both programs of the YWCA of Troy-Cohoes.

Influenced by my work in community technology centers and the policy rhetoric popular at the time, I initiated a project designed to close the digital divide by providing situated technology training, asset-based community development, and workforce preparation for low-income women. But women in the YWCA community repeatedly disputed and disrupted the digital divide frame. As my relationships with them developed, they described their struggles to meet their basic needs in the high-tech economy and their significant, often troubling, interactions with the tools of the information revolution. When given the opportunity, my collaborators even smashed the machines. Literally. Gleefully.

"If I Had a Magic Wand, I Would Bomb All the Fucking Computers"

When I started interviewing women at the YWCA of Troy-Cohoes in the summer of 2003, I had been working in the YW community for nearly two years. One interview, with Ruth Delgado Guzman, exemplifies the challenges women in the YWCA community posed to digital divide framings and begins to illustrate how their insights shifted my understanding of high-tech equity. Ruth and I met through the Women's Economic Empowerment Series, a nine-part sequence of popular education workshops that I co-designed and facilitated with YWCA staff member Christine Nealon during the summer of 2002. Ruth later became a member of Women at the YWCA Making Social Movement (WYMSM), our collaborative research and social justice group.

Ruth is an engaging, eloquent Puerto Rican woman who was completing her master's degree in education at Russell Sage College in Troy at the time. Deeply committed to the well-being of children and hoping to become a high school guidance counselor, she kept me laughing with her sharp wit and inspired me with her abiding interest in social justice. Our interview took place in late July. The windows of the Sally Catlin Resource Center were open over State Street, and the center was flooded with light. Oscillating fans worked to move the warm air in the room as we talked about information technology (IT) and social justice. Other members of the YW community typed quietly on the public computers, and the intercom cut in and out of our conversation, announcing phone calls and visitors for residents and staff. Ruth was quick to name the goal she felt we both shared: creating a "technology for people." She described her experiences with technology as generally "very, very positive," and explained that she believed very strongly that technology could be used as a tool of social change.

However, she expressed reservations about most scholarship that describes women's technological inequality, and about the public policy geared toward alleviating it. She insisted,

People who say that women are afraid of technology, or don't know how important it is, are missing the point. . . . When you're just surviving, you're in survival mode. You don't think about technology, you don't think about the latest anything. You are surviving. And that takes your whole life—just to survive. . . . Especially women! Women love to learn and are able to learn. They really like technology and want technology. If you offered women a system that they created, for everyone, they would want it, they would engage with it. But it's not like that.

Computers, software, and Internet architecture are designed for financial people and for business people, for professionals, she argued. "But where are the mothers," she asked, "or people who work and struggle to stay afloat? The homeless?" Digital divide policy, she insisted, does not address social and economic justice issues central to the lives of people who struggle to meet their basic needs. "It's not technology that will make our

Box 1.1
WYMSM Member Profile, Ruth Delgado Guzman

"If you find the courage to keep going, that changes everything."

I am a fighter. I have resiliency. I like to improve myself. What I learn in books, in theory, I put it into practice. I work for my dreams.

I ended up in the YWCA because I had a breakdown, emotionally. I was doing my master's degree and I was by myself in the residence halls, with no family nearby. I had to stop taking a few of my classes because I couldn't handle it. That's why I moved to the YW: to be in the residence you have to have at least nine credits. The move was devastating. It hurt my self-esteem, it was demeaning, but I wanted to keep going with my studies. The move to the YWCA was tough. There is a stigma about living there. There is also a stigma about emotional problems if you face them and say something about it. Being a minority also.

WYMSM helped me stay in school. You guys helped a lot. [Not everyone in WYMSM] lived at the YW, or went through what I went through, but we were all intellectually at the same level. Sometimes I feel like I don't fit—I'm a square peg in a round hole—so for me, being able to relate to people like me was very important. To be able to see you, and hug you. Warmth! Human warmth. There were so many things: drum circle, sitting down to discuss things, having coffee, HTML classes. I could ask you things I couldn't ask anyone else. So I related to you, I developed rapport with you, and that helped me.

The YWCA helped me to grow and have more depth, because I saw situations that you don't normally see, when you are in the bubble of your family or your limited experience. If I hadn't been living at the YWCA, maybe I would not have gotten involved with the things that I did. WYMSM helped me grow emotionally, physically, spiritually. It was holistic in a way that I was able to say, "Oh, my God. I can do it. I am resilient." I kept going because there were other people I could help. I was not the only one who was suffering. As a matter of fact, there were others who were worse off than me. It made me realize that I could share myself. That helped me.

Five years from now, I want a good job. I miss the drum circles that I used to do at the YWCA, and the group therapy I did for different kinds of people. I am a school counselor, and I didn't do that kind of work in the schools. Eventually, I would like to have my own clinic where we will mix rhythm and music with emotional and spiritual growth. I would like to keep helping people. But to do that I need financial means, which is why I need the job. My relationship is also one of the things that I'd like to have together in five years.

Whatever circumstances you have in your life, if you find the courage to keep going that changes everything. You will have a sense of accomplishment. We all have a lot of potential, but sometimes we don't know it and don't tap into it, because of self-esteem or abuse, or whatever we go through. We have to learn from it and make a commitment to keep learning and growing.

Based on a phone conversation that took place on May 31, 2009.

lives better. That will make us 'haves,'" she explained. "It's social conditions, financial conditions, the environment. Technology is just a little part
of it . . . it's not *justice*."

Technology for people would be different from universal access to existing computer systems, she argued. It would mix systems "designed by
women, for everybody" with educational programs combining functional
goals such as finding housing with technology skills training such as Web
searching in order to increase people's well-being financially, emotionally,
socially, and intellectually. She joined WYMSM primarily to have opportunities to brainstorm about building such systems.

Community technology centers around the country and the world have
gone a long way toward fulfilling Ruth's vision of a technology for people
by tirelessly wiring communities, providing them with affordable access
to information and communication resources, nurturing generations of
trainer-activists, and preparing the ground on which community-produced
content can grow. This is powerful, crucial social justice work performed
by committed and innovative people and organizations. But these efforts
are primarily redistributive, focused on providing access to the tools of the
information revolution to communities that, it is assumed, lack such
access. This assumption has led some to characterize low-income communities as technology-poor.

But people struggling to meet their basic needs do *not* lack interaction
with IT. As I describe throughout this book, if women in the YWCA community are any indication, poor and working-class women have a great
deal of interaction with IT in their everyday lives, particularly in the low-
wage workforce and the social service system. The assumption that poor
and working-class people lack access to technology, broadly generalized,
has led to policy and community organizing approaches that are practically misguided. In solely redistributive schemes, marginalized communities—and the people who live in them—are seen only as products of lack
and destitution, not as vast reservoirs of assets, resources, networks, expertise, strength, hope, passion, and innovation. The assumption of community deficit blinds many policymakers and community organizers to the
real world of IT, to the true relationship between technology and poverty,
and to the hope for high-tech equity.

I admit that when I started, I held this misconception too. I came to
this project convinced that distributive approaches were the linchpin of a
more just information age. As a committed community technology practitioner, I used my skills to increase access and teach technical proficiency
to close the digital divide. But women in the YWCA community routinely

challenged my assumptions, both implicitly and explicitly, and over the course of my first two years in the community, I slowly realized that I was laboring under false pretenses. One class we offered at the YWCA, "How Does the D@mn Thing Work?," provides a poignant example of why I was forced to reconsider my presumptions.

The workshop was loosely structured around the demolition of an unusable donated computer.[2] We took the cover off the machine, handed out screwdrivers, told participants that it was going into the Dumpster anyway, and then let them do whatever they wanted to it. As parts came out of the computer, we passed them around and told everybody what each part did. For a few minutes, women carefully extracted networking cards and hard drives from the PC and gingerly examined them. After a bit of time and some convincing—they were particularly concerned about waste, wanting to be absolutely sure that no one in the YW or elsewhere could use the computer before they took it apart—they started to believe they could do whatever they wanted to it, and the demolition began. They hacked at the computer. Broke pieces off, and then broke them into smaller pieces. Put the pieces on the floor and jumped on them. Tore apart bundles of wires, wedged off covers to see the motors and chips—all with a palpable sense of glee.

What was I to make of this merry destruction? How was I to reconcile the tangible wave of frustration that set off the demolition of the computer with the hope and optimism expressed by Ruth Delgado Guzman for building a technology for people? When I began to do interviews a year later, I probed women in the YWCA community about their everyday experiences with IT. Where, I asked, did they come into contact with IT in their daily lives? What were those experiences like? Their answers were surprising. Some women certainly responded in ways digital divide scholars and policymakers would have predicted: they spoke at length about the inequitable distribution of technology, declared their desire for better access, and described their day-to-day use of IT to find important information and support their social networks.

But the majority of women I interviewed in the YWCA community talked about a different kind of experience with technology altogether, an experience marked not by technology lack or deficiency but by technological *ubiquity*. They described their extensive use of computers in the low-wage workforce—about half of the women I interviewed had been data entry or call center workers. Others talked about encountering computers in the social service system. They described welfare caseworkers who blocked eye contact by hiding behind a computer terminal. They described their feelings of hopelessness and frustration when caseworkers couldn't find their information in "the system," a feeling intensified by the computer's apparent power to decide their family's fate. In their experiences with the databases of Medicaid or Social Security, many described feeling that they "became a number," and complained that the computers "find out everything about you." Surveillance technologies such as metal detectors and fingerprint machines, they argued, turn public agencies such as schools and social service offices into prisons. Several voiced suspicions that regional initiatives like the local Tech Valley development were going to result in higher rents in exchange for a few jobs for the college-educated, benefiting a small percentage of area residents and disadvantaging many. These concerns are broadly structural and impossible to address by redistributing IT or providing access to computers.

The role technology played in the lives of women in the YWCA community was characterized by ambivalence, not absence. They were optimistic about technology's potential but concerned about its real-world impacts. They expressed strong conflicting emotions in interviews and at public events: hope for a better future, excitement about new innovations, anger over continuing injustice, and cynicism about efforts that used technology to alleviate that inequality. Smashing the computer could be read, I realized, as an effect of the extraordinarily complicated relationship women in the YWCA community had with technology, even as an attempt to take power back from a symbol of the system. In a later interview, Veronica Macey, a participant in the workshop, confirmed this interpretation:

That taking apart the computer thing really helped [engage women who feel out of the technology loop]. Because I know I never saw [Patty] at any computer type stuff before and that seemed to help her get into it. . . . What's in the inside guts? I can break it apart! It's not this big scary thing, I can *kick* it and things come off. That helped. Stuff like that that shows that computers are not these big infallible immortal objects.

Initially, I thought that the computer-smashing incident was a quirk, an outlier. But evidence of the ambivalence of women in the YWCA community continued to mount. For example, they would engage in technology training courses meant to prepare them for the high-tech workplace while strongly expressing their doubts that the training would in fact lead to a sustainable job. Even now, after my time at the YWCA, I continue to experience poor and working-class women's ambivalence in the face of IT. Engaged in a new project about the citizenship impacts of welfare administration technology, I was recently talking to an interviewee who had faced many struggles accessing public assistance, among them a series of technical glitches. Near the end of the interview, I asked her, if she had a magic wand, what one thing she would do to change how technology is used in the social service system. She replied, "If I had a magic wand, I would bomb all the fucking computers."

Women in the YWCA community were quite reasonably conflicted about their relationship with technologies that are simultaneously symbols of knowledge, power, and opportunity and instruments of their surveillance, discipline, exploitation, and oppression. Critical ambivalence is a sign of incipient analysis:[3] women in the YWCA community were noting the mismatch between the image of computers as the route to social and economic progress and their own experience of the technology as intrusive and limiting. Too often, policymakers and scholars misread this ambivalence as "reluctance" or "inability" to engage with technology and technological training rather than seeing it as the ground from which a critical consciousness about the relationship between technology and inequity grows. It was not until I got past my own class- and race-based assumptions about technology that I began to understand this critical ambivalence and use it as a resource in our efforts to make IT a tool for social change.[4]

Life in the State of Poverty

On March 21, 2002, WYMSM held its first major public event, an open forum recognizing Hunger Awareness Day during which community members were invited to try to survive one month on a Rensselaer County welfare check. The event, which used a State of Poverty simulation exercise designed by Missouri-based Reform Organization of Welfare (ROWEL),[5] occurred in the gymnasium of the YWCA. On the walls and on refrigerator boxes, student interns, residents, staff, community members, and their children had painted a cardboard city. At tables that ringed the room sat residents of the YWCA—resourceful women, young and old, African American, Latina, and white, native-born and immigrant, mostly struggling to meet their basic needs—prepared to play roles they were intimately familiar with as customers and clients. For this day, they would experience these roles from the other side of the desk. They would be the bankers, pawnshop owners, welfare caseworkers, teachers, and police. They would be the politicians and power brokers; they would run community services and local businesses. One participant was even charged with running

Resourceful Women
FIGHT HUNGER

A Simulation of the State of Poverty in honor of
H u n g e r   A w a r e n e s s   D a y

Hunger Awareness Day is a day to call for an end to hunger
and its causes. In its honor, the YWCA of Troy-Cohoes is
hosting ROWEL's "State of Poverty" welfare simulation to
educate legislators, city officials, students, and other
community members about what it takes for low-income
individuals and families on welfare to survive from month to
month.

Please join us for this
fun and fascinating
public event!

Thursday, March 21 3 - 6 PM
The YWCA of Troy-Cohoes
21 First Street (corner of First and State streets), Troy
For more information, call the YWCA at 274-7100

Hosted by Women at the YWCA Making Social Movements (WYMSM), RPI's Community
Outreach Partnership Center (COPC), and the Hunger Awareness Day Steering Committee

Figure 1.1
Hunger Awareness Day event flyer.

"illegal activities" for the town. Together, we were prepared to create an imaginary community for the public to inhabit for a single hour, in an effort to highlight the ongoing struggles of families trying to survive the always stingy and often unjust system of public assistance.

We had worked since January to plan the event, in collaboration with Hunger Action Network of New York (HANNYS), Statewide Emergency Network for Social and Economic Security (SENSES), and students and faculty from Rensselaer Polytechnic Institute. After a series of two workshops held at the YWCA under the title of Women, Simulation and Social Change, a core group of academics, YWCA residents, YWCA staff members, and community members emerged to form the collaborative group, WYMSM.[6] The group was central to the planning of the Hunger Awareness Day event: we helped write the press releases, recruited participants, gathered props, spoke at the press conference, and prepared food for a community meal to follow the event. Most important for our values and mission as a group, we spent considerable time recruiting community members who had actually experienced poverty—women we considered the real experts—to play community resource provider roles so that they could educate the general public about their experiences.[7]

Moments before our start time, Jes and Christine stood at the doors with the sign-in sheets; Jenn, Patty, and Coffee held notes for their speeches; community resource people sat at their tables and gathered their paperwork and props; documenters checked the battery levels in their video cameras and set their white balances; I stood in the middle of the room with my clipboard and my whistle. And we waited. The first big group to arrive was from Sand Lake Baptist Church, mostly women, mostly in their seventies and eighties. Then grade-school students, ten- and eleven-year-olds from the Susan Odell Taylor School; then members of the Ironweed Collective, an anarchist group from nearby Albany. Then people just streamed in. We had no idea who they were, but more than ninety people came.

We gathered the crowd briefly for a press conference. A representative from a statewide antihunger organization laid out recent changes to welfare and how they would affect hunger in New York. Rensselaer faculty and WYMSM member Nancy D. Campbell spoke of the need to build relationships between communities and universities. Executive Director Pat Dinkelaker spoke about the rich community of the YWCA of Troy-Cohoes, describing how the organization was building a grassroots movement to fight hunger in the Capital Region. Finally, it was Coffee, Patty, and Jenn's turn. They described their experiences, Coffee quickly, Patty in a shaking

Figure 1.2
Women, Simulation, and Social Change workshop, November 12, 2001. Visible are, clockwise: Chitsuge (Chris) Mapondera, Virginia Eubanks (laughing in back), Patty Marshall, Jennifer Rose (at computer).
Photo: Pat Dinkelaker

Women, Simulation, & Social Change *Working Group*

We're seeking volunteer staffers to help run the ROWEL "Life in the State of Poverty" simulation game to honor Hunger Awareness Day at the YWCA on March 21, 2002.

Are you interested in promoting social change and fighting poverty?

Do you have an afternoon to spend educating Troy public officials and citizens?

If so, you're invited to a planning workshop:

Women, Simulation, and Social Change

This workshop will lead to opportunities to work on a media autobiography project in collaboration with the RPI Community Outreach Partnership Center in the future.

Discussion! Doorprizes! Animal Noises!

Figure 1.3
Women, Simulation and Social Change working group flyer. WYMSM grew out of this initial group.

voice, and Jenn speaking from her heart without looking at her prepared notes. "Maybe," she said of the simulation that the public was about to undertake, "if everyone experienced this, even for an hour, maybe we could get together and we could *do* something. Make change."

After the press conference, I welcomed the public and explained that the object of the simulation was to sensitize participants to the day-to-day realities of life faced by poor and working-class people and to motivate them to organize to reduce poverty in the United States. I reminded them that the exercise was a simulation, not a game: the statistics and situations used were accurate, based on real-life experiences of low-income families.

As quickly as possible, I split attendees up into twenty-six families—single moms and their children, elderly couples, and nuclear families—and gave them a packet of information to read about their roles, their resources, and the goal of the simulation: provide for your family's basic needs for one month. We asked them to be as realistic as possible about the roles they were taking on, introduced them to community resources (the landlord, police, bank, pawnshop, food bank, employment office, social services, etc.), and I blew my whistle to start the first of four fifteen-minute "weeks" of life in the state of poverty.

At the end of the one-hour month, we gathered together to discuss the experience. In a big circle, participants reflected on their attempts to fulfill their basic needs: obtaining or keeping shelter; maintaining utilities, gas and electric; buying food; making loan payments; keeping their children in school. They told stories about what had happened to their families, speaking with great emotion about how community resource people responded to their needs (or didn't). One woman complained that the landlord had taken her rent money, but because she hadn't asked for a receipt, the landlord reported her to the police as not having paid, and she was evicted. Others protested that the lack of good community resources, such as affordable child care or reliable transportation, kept them from meeting their goals despite Herculean efforts. One participant said she found the experience so frustrating she was nearly moved to strike her simulated "wife." A YWCA community member playing a social service worker described her struggles to help participants navigate the system, which ended too soon when she was "shot" by the illegal activities person in an attempted robbery.

It was not all hardship and frustration. Participants helped each other, trading extra resources among families, trying to succeed through mutual aid. Pat Dinkelaker snuck in during the event and started organizing participants into simulated social movements. Halfway through the discussion, they stood up en masse and declared their intention to end poverty in the real world—by fighting to reform welfare, advocating for affordable day care, organizing to raise the minimum wage. Women from the YWCA community who played roles as community resource people spoke eloquently about how realistic the simulation was and how much it meant to them to be able to help members of the general public understand their experiences and develop a more accurate picture of poverty in our community.[8]

At a WYMSM meeting a week later, we discussed our experiences and described our favorite moments. All of us were proud that we had

Figure 1.4
Hunger Awareness Day discussion.
Photo: Pat Dinkelaker

pulled off this enormous event. Coffee said, "After my speech . . . I felt good about myself, I felt really good. I didn't know I could do that. . . . That's a powerful thing, you know what I mean? That you can do these things. You can plan to do them and do it right." Coffee, Patty, and Jenn each mentioned speaking at the press conference as their proudest moment of the event. But when answering the question, "What was the most profound experience you had?," nearly everyone in the group mentioned moments when new understandings were reached, new empathies formed across the barriers of difference. For example,

Jenn: The most profound thing I experienced at HAD was . . . listening to one of the kids from school, at the end when he spoke. He couldn't have been more than ten, but it was really amazing. It was like, whoa!
Virginia: Do you remember what he said?
Coffee: He kind of said, the way he lives, he figured [that] was how everybody else lives. Things are different [for other people], and he didn't know that. . . . He thought everyone was like [him].
Jes: [That] little boy said that he never knew what it was like. He was like, "*This worked.*" Hopefully it will change at least the way *he* thinks because of how he's been raised. I was like, "Yeah!"

The Hunger Awareness Day event demonstrated some of the rewards of participatory projects: the energy and impact of consciously structured collective process, the power of voicing your own experiences and joining with others to reshape your world. It hinted at the complicated relationship between personal experience and expertise in mass mobilization, and gestured toward the transformative power of experiencing information firsthand rather than just reading about it. The event also illustrated some of the challenges of participatory work. Participatory projects, even those that are committed to ending poverty, often ignore or marginalize the voices of poor and working-class women. A sustained commitment to social justice organizing can be hard to uphold in the face of people's basic, pressing needs. Forming relationships across the lines of race, class, and gender is difficult. But despite these struggles, WYMSM agreed that there was a moment, standing in the circle at the end of the simulation, when everything just clicked, when people talked honestly and intelligently about power, privilege, and poverty. A moment when we all had hope.

This Is a Love Story . . .

Remembering that moment, and honoring it, means that this book is also a love story about collective process. Many people fall in love with collective activity—social movements, mass mobilizations, encounter groups, cults, even gathering in the neighborhood bar. The flush of recognition and belonging, the heady erotic charge, and the comfort of unified action are as much a part of joining a social movement as they are symptoms of falling in love with an individual. As romantic mainstays, "together at the barricades" or "fighting together against all odds" are almost as popular as "love at first sight." Falling in love with a practice, like falling in love with a person, is easy. It is staying in love, the ongoing act of loving rather than the attainment of the love object, that is hard. This is particularly the case when the parties involved are deeply dissimilar, when love reaches across the boundaries of difference or challenges fiercely held social norms. Maybe this is why so many stories of star-crossed love end in death.

Sometimes both the lover and the beloved die—as in *Romeo and Juliet*—but usually it is the free spirit, the boundary transgressor, the social upstart, the poor one. It is almost always the woman who is sacrificed to liberate her male lover. In *Love Story*, the preppy lives, never having to say he's sorry, and the scholarship girl who loved above her station succumbs. In *Harold and Maude*, free-spirited, car-stealing septuagenarian Ruth Gordon teaches cloistered, bored, death-obsessed heir Bud Cort to live and love more fully, and then commits suicide. More recently, Wynona Rider and Charlize Theron taught Richard Gere and Keanu Reeves, respectively, to cherish every moment before conveniently perishing, in *August in New York* and the 2001 remake of *Sweet November*. As someone who believes in the transformative potential of social movements and the limitless power of love, I find the popularity of this narrative—the impossibility of sustaining passionate commitment across difference—troubling.

As a feminist, I also find the pile of dead girls disturbing. Like the character that Spike Lee calls the "magical Negro,"[9] whose special powers exist to get a white protagonist out of trouble or to teach him about his faults, the "sacrificial girl" in these films is an object only, a means to the end of increased self-knowledge for the male protagonist. After he realizes his own deficiencies, what use is the girl? If she doesn't quickly get out of the way, the cameras fade to black as soon as the improbable match is made—think *Pretty in Pink* or *Pretty Woman*. The mundane but crucial questions of maintaining a relationship across difference are never explored. When they go out, who pays? How do they deal with their families during the holidays? Raise their kids?

Similarly, we have a million stories about diverse groups of people uniting in a final showdown against evil; the entire *Star Wars* franchise is an obvious example. But we do not have many stories that show us how hard, and how rewarding, it is to *actually* forge and maintain alliances across difference. We don't often see the realities of the three-hour meetings, the lost opportunities, the hurt feelings, the passionate misunderstandings and the day-to-day difficulties of organizing across class, race, gender, nation, and sexuality. As with star-crossed love, even when we dare to believe that organizing across difference is possible, we seem to find the details terribly boring. Easier just to posit some mythical "Movement moment" when differences are put aside, to deify a superhuman charismatic leader who turns divisiveness into coalition, or to mourn the sacrificial lambs, who become a rallying cry for unity. Easier to ignore or forget the day-to-day heroism of ordinary people coming together to transform their world.

Perhaps a love story is a lot less romantic if we deal with the details, the practice of loving, rather than the climactic plot point of falling in love. But love isn't something you "fall into." Love is an action, something you do, a choice you make in every moment. The romantic story—the one without the details—offers an impoverished model of love, one that relies only on luck, chemistry, and short-lived and often one-sided sacrifice. This unattainable myth misrepresents both loving practice and collective process; it is a pale imitation of what Dr. Martin Luther King Jr. called "the long and bitter, but beautiful struggle for a new world."[10] We should demand more.

Figure 1.5
A closing exercise in the Women's Economic Empowerment Series—knitting participants together. Left to right: Cosandra Jennings, Ruth Delgado Guzman, Jenn Rose, Nancy D. Campbell.
Photo: Pat Dinkelaker

Actual struggles to build relationships across difference are both fascinating and inspiring. In this book, I try to capture some of the everyday triumphs and failures that followed from trying to use technology as a tool of social change in a small city in upstate New York. This book is a love story, but it is not a very romantic one. Over time, our collective work in WYMSM became an attempt to create a love story with no "sacrificial girl," to build and sustain relationships between free people engaged in meaningful struggle in the real world. The process demanded that we conquer fear, build alliances, and try to speak our deepest truths. The story I tell here is about taking risks and creating spaces where vulnerability results in transformation rather than in physical and emotional harm.[11]

I am sure that a story about collective social transformation is not what you expected from a book about the profoundly rational world of computers, public policy, and high-tech economic development. That may be because we tend to think of technology as a destiny, not a scene of struggle; a product, not a site of possibility; a static, ahistoric thing, not "an ambivalent process of development suspended between different possibilities" (Feenberg 1991, 14). We often see technology as only an object, cut off from social relations, floating in space like HAL in *2001: A Space Odyssey*. But technology embodies human relationships, legislates behavior, and shapes citizenship. Our mistaken assumptions about technology's static "thingness" prevent us from recognizing the real world of IT, and from realizing what Ruth called "technology for people."

2 The Real World of Information Technology

Central to any new order that can shape and direct technology and human destiny will be a renewed emphasis on the concept of justice.

—Ursula Franklin, *The Real World of Technology* (1999, 5)

This is not a book about the digital divide. The relationship between inequality and information technology (IT) is far more complex than any picture portraying "haves" and "have-nots" can represent. Working toward an information age that protects human rights and acknowledges human dignity is far more difficult than strategies centered on access and technology distribution allow. One piece of the high-tech equity puzzle that is generally overlooked when we try to imagine "technology for people" is the relationship among technology, citizenship, and social justice. This is unfortunate, as our notions of governance, identity, and political demand making are deeply influenced by IT in a wide variety of institutions, including social service agencies, training programs, schools and colleges, government institutions, community organizations, the workplace, and the home.

In my own life, IT seems to fulfill the high-tech economy boosters' most ambitious promises. It enriches the democratic process and expands my opportunities for a fulfilling and prosperous career. Online databases and Web-based portals ease my relationship with government, increase transparency and accountability, and simplify my daily tasks. I can register my truck online in an instant, pay my taxes with a nifty computer program, or check stimulus package spending though the Web. In my workplace and professional networks, digital technologies facilitate mobility and multiply opportunities, allowing me to undertake many job tasks, such as planning courses, advising students, or communicating with colleagues, from any location. But, as I argue throughout this book, where people are located in relationship to the power structures of society—that is, their "social location"—has an enormous impact on how they encounter and relate to the tools of the information revolution.

Women in the YWCA community have a more complicated relationship to IT. In interviews and public events, many have said that they think "computers are the future," and have conceded that technological skills are something they should have to remain competitive in the job market. But they also directly experience the more exploitative face of IT as workers in low-wage, high-tech occupations such as data entry and call centers, as clients of increasingly computerized government services, and as citizens surveyed by technologies in public institutions and spaces. It is not so much that they lack access to technology but that their everyday experiences with it can be invasive, intrusive, and extractive.

As I began this research, my overreliance on access-based models of high-tech equity and my privileged class and racial position led me to overlook—and even obscure—poor and working-class women's extensive interaction with information systems. In many cases, rather than being "technology poor," women in the YWCA community had *too much interaction* and *too intimate a relationship* with IT. They described living as subjects of far-reaching technological regimes, not as technological "have-nots." Spurred by their insights, I began to study high-tech equity differently. What can we learn about technology, poverty, governance, and resistance, I asked myself, if we actually attend to the everyday experiences of women in the YWCA community?

Over time, through collaborative research and education programs at the YWCA, we unearthed the exclusions, obfuscations, and denials at the heart of so much policymaking around high-tech equity. The participatory approaches we employed revealed data and interpretations that called into question many long-held beliefs of technology policymakers and scholars: that low-income people are technology poor, that technology training leads to sustainable employment, that women are more reluctant than men to engage with complex technological systems. But how could the mainstream policy community get it so wrong?

I believe there are five main problems that create a myopic focus on access in technology policy:

1. Our ideas of *equity* and *justice* are too limited. We are trapped in a distributive paradigm that understands high-tech equity only in terms of the availability of information products and resources.
2. We ignore feminist insights about the impact of *social location* on all people's experiences with the tools of the information revolution. An

intersectional approach to studying social location and everyday life makes the complex inequalities of the information age visible.

3. Our understandings of *citizenship* are too narrow. We rely on limited, liberal humanist conceptions of citizenship that obscure the operation of technology to teach lessons about political participation and power.

4. Our definitions of *technology* are too static. We think of technology as a set of artifacts, not as an assembly of practices for organizing the world that encode some norms, values, and ways of life at the expense of others.

5. Our *methodology* is neither adequately rigorous nor sufficiently modest. We too often go it alone, neglecting broadly participatory research and program design that can bring badly needed empirical specificity to our studies of and interventions in the inequalities of the information age.

In this chapter I explore each of these barriers to acknowledging the technological experiences of women in the YWCA community and to understanding the relationship between technology and women's poverty more generally.

Equity and Justice: Displacing the Distributive Paradigm

Mainstream policymakers in the United States often conflate equity or justice with fair distribution. This is not entirely surprising. American political culture is strongly attached to the idea of citizenship as a contract between individuals and the state, an arrangement that consists of rights and responsibilities, stimulates personal accomplishment, and protects individual liberties. If citizenship is a contract, then each of us as an individual free agent is responsible for getting the best deal we can for ourselves, and it is the government's job to lower any preexisting or structural barriers to our efforts. Our mainstream idea of justice is thus captured by the image of a level playing field, a picture that relies on understanding justice as fair distribution of opportunity.

Embedded in this context, most work on technology and social justice has focused on the challenges and benefits of providing marginalized populations access to high-tech tools. Following from this "distributional ethic" (Cozzens 2007), policy solutions tend to center on fairer distribution of the risks, benefits, and products of science and technology among individuals in society. There are certainly pressing reasons for such a focus: in times of growing economic catastrophe, the allocation of material goods, including the products of technoscientific research and development, should be on any social justice agenda.

However, for nearly twenty years, feminist political theorists have called into question the overreliance on fair distribution as the primary route to justice.[1] For example, Iris Marion Young writes that there is a "distributive paradigm" in contemporary political theory that defines social justice too narrowly "as the morally proper distribution of social benefits and burdens among society's members" (Young 1990, 18). According to Young, the goods doled out in redistributive schemes generally include wealth, income, and other material resources, but the concept is often stretched to include "nonmaterial social goods" such as rights, opportunity, power, and self-respect.

Young argues that public policy produced through the framework of the distributive paradigm is flawed on several counts. First, it places too great an emphasis on fair division, ignoring other social values and obscuring the role of institutional context and social structure in perpetuating injustice. It also presupposes social atomism, leading scholars and policymakers to focus on the individual agent and to miss the role that social groups and social structure play in perpetuating oppression and domination.[2] The distributive paradigm's focus on allotment leads scholars and policymakers to focus on end-states rather than on processes. Policy created through the distributive paradigm misrepresents nonmaterial social goods—rights, respect, dignity, power—when it is applied to them (ibid., 18–31). Finally, Young argues that critical concerns that are not of a distributive nature— freedom from cultural imperialism, transparent political processes, safe living and working environments, ending gender-based violence—are ignored in exclusively distributional models of justice.

Focusing our high-tech equity efforts solely on distribution—on equal access to technology—creates similar distortions. Seeing high-tech equity *only* as broadly shared access to existing technological products ignores other social values, neglects decision-making processes, sees citizens only as consumers, and ignores the operation of institutions and social structure. The myopic focus on access is part of a distributive paradigm in IT policy that sees all high-tech equity issues as redistributive issues. Although access is an important part of the high-tech equity puzzle, it is not adequate for developing strategies to combat inequalities of the information age that are not material or distributional, such as cultural recognition, institutional discrimination, health and safety issues, environmental injustice, nonparticipatory or ambiguous decision-making structures, and rights to privacy and technological due process.[3] The distributive paradigm, with its focus on equity as remedying perceived citizen deficits, is a weak foundation on which to build a less oppressive, less exploitative technological present.

If we displace the distributive paradigm, we can develop a model of high-tech equity based on resisting oppression, acknowledging difference as a resource, and fostering democratic and participatory decision making.

Social Location: Grounding the Information Revolution

Feminist scholars have offered important insights into the relationships among science, technology, and society.[4] But there are crucial gaps in feminist social studies of IT that limit the usefulness of this scholarship in addressing broader concerns about technology, citizenship, and social justice. Most feminist analysis of IT has centered on three issues: cyberspace, cyborgs, and pipelines. Feminist work on cyberspace tends to explore discursive practices online and the representation of women in informational contexts.[5] Feminists theorizing the cyborg[6]—a hybrid figure that embodies contemporary anxiety about the crossing of boundaries between animal, human, and technology—concentrate on the interface of women's bodies and machines, particularly in the realm of biomedical technologies, including new reproductive technologies. What dominates most feminist empirical work on IT, though, is concern about the pipeline,[7] the path to high-status occupations in science, technology, engineering, and mathematics. Most analyses of the failure of women and men of color to fill the gap in the pipeline focus solely on highly paid technological occupations such as tenured faculty positions at research universities or managerial positions in high-tech industries. This overlooks the significant role low-wage workers in a wide variety of jobs—such as industrial and product assembly, packaging and delivery, customer service, data entry, and the caring occupations—play in the high-tech economy.

Cyberspace, cyborgs, and pipelines are certainly interesting and worthy of study. But feminist analyses of IT in these areas tend to embed assumptions about possessive individualist citizenship, privilege the experiences of wealthy and educationally advantaged people in the global North, and view technology as merely a consumer product, access to which is an unmitigated social good. Too often, in our attempts to break into the boys' club of science and technology, feminists have collapsed all vectors of difference down to a single marker, gender. Thus, most feminist analyses of high-tech equity are captured by the field of "women and technology" which spotlights the absence of women in technology research, design, and use. But this approach to high-tech equity centers the experiences of relatively privileged women, which makes it difficult to understand and account for the experiences of women who inhabit vastly different social locations—and who thus have vastly different experiences with IT—because of their class, ethnicity, race, sexuality, ability, and nationality.

In addition, "women and technology" approaches tend to discount the vast experience most women have with IT by defining users as those who own a personal computer or have access to one in a community center or library and by defining a high-tech worker as someone who has gained entry into the high-status, high-paid world of the knowledge industries. But what about the woman who encounters the back of a computer terminal or a digitized voice on the phone instead of a caseworker when trying to access public benefits? Or the woman working contingent, low-paying, part-time jobs in data entry and call centers? Or the woman half a world away who assembles the circuit boards, monitors, and keyboards that bring the information revolution to the United States? Or the woman whose family disassembles these same devices to recover precious metals several years later with no protective gear, environmental controls, or labor laws?

These are significant, everyday experiences of the information age that are shared by millions of women the world over. These are, in fact, *the great majority* of women's experiences with IT, both abroad and at home. Too often, even in the feminist literature, what counts as "real" interaction with IT is defined in profoundly class-, race-, and nation-specific ways. Conceiving of women's relationship with technology in this limited manner obscures the experiences of women marginalized by their class, race, sexuality, nationality, ability, or age; ignores their concerns; forecloses their knowledge and energy; and often concludes that they are inevitable, if unfortunate, victims of progress. This is a serious limitation in our work as feminist critics of science and technology, one that constrains our ability to push science and technology toward social justice.

In contrast, intersectional approaches to studying how social location shapes *all* women's experience with technology in their everyday lives make the structural violence of the information age visible. Rendering oppression visible makes it available for intervention and change. Intersectional approaches acknowledge the various ways that gender, race, class, and other categories of difference interconnect to create a particular social location from which each woman—and each group of women— experiences everyday life, including interactions with technology. Without the careful attention to social location required by intersectional approaches, cyberspace remains a fictive friction-free space for identity play, cut loose from bothersome "ism" associations such as race or class. Without attention to how race and class shape women's technological choices, the cyborg is seen only as liberating women from the tyranny of their bodies, and therefore from their specific location in social relations. Without careful scrutiny of how women, especially women struggling to meet their basic needs, actually interact with technology in their everyday lives, it is too easy to mourn their absence in the pipeline.

Through the intersectional lens of social location, cyberspace, cyborgs, and pipelines look very different. It becomes obvious that larger cultural and political conversations about race, class, and gender mediate and create discursive spaces on the Internet.[8] Cyborg interventions such as surrogate pregnancy, which liberate wealthy women and enable them to increase control over their reproductive decisions, can work to commodify or limit poor and working-class women's reproductive choices, dehumanize and demonize their motherhood, and erode their self-determination and health.[9] Through an intersectional lens, it becomes obvious that a woman's experience of the information economy is very much dependent on where she stands in relation to power. Wealthy women's increasing entry into high-tech, high-status occupations—as well as elite men's ability to continue to neglect the work of caregiving—relies extensively on battalions of women working jobs in service, caregiving, and low-status IT occupations.[10] Low-paid, exploitative and contingent work by women all over the world holds up the pipeline for a few of their privileged sisters.

Studies focusing on women's involvement in the design and use of IT tools, the high-tech workforce, e-governance, e-commerce and business networks, or as consumers of information are important, but most ignore and marginalize intragroup differences. It is just these differences—particularly the growing inequality among women by race and class—that define injustice in the information age.[11] Intersectional approaches make high-tech equity issues visible that single-issue, "women and technology" analyses obscure: the superexploited labor that provides the scaffolding for "friction-free" capitalism, the high-tech state surveillance that creates manageable citizen-subjects for neoliberalism, and myriad forms of resistance surrounding, and mediated through, IT.

Citizenship: Fostering Reflective Technological Citizens

Technology is a way of shaping what it means to govern and what it means to be a citizen.[12] Citizenship activities such as voting or claiming welfare benefits are increasingly routed through complex and often poorly understood IT applications. Interaction with IT in state bureaucracies and government institutions provides an opportunity for political learning, teaching lessons about social worth, competence, opportunities, rights, responsibilities, and risks. These technologies—including management information systems, electronic benefits transfer systems, closed-circuit television surveillance systems, and biometrics systems such as automatic fingerprint identification, hand geometry, and retinal scanning—help create different modes and forms of democratic citizenship. They can be coercive or liberating, depending on one's social location and the processes of the technology's design and implementation.

IT can complicate the relationship between citizens and the institutions that affect their lives. For example, one of the great promises made about the integration of IT into the processes of governance was increased transparency and accountability.[13] But the burden of increased visibility has fallen almost exclusively on America's poorest families. Poor and working-class women, especially those attempting to access public assistance, became hypervisible to government agencies as administrative IT applications exploded in the last decade in the public assistance, criminal justice, and educational systems. This visibility can increase their physical and economic vulnerability, because heightened monitoring of poor and work-ing-class families results in the discovery of behavior that "would have gone unnoticed had it occurred in the privacy afforded wealthier families" (Roberts 2003, 32). Such monitoring in turn creates a relationship between citizen and state based on mistrust, deception, and risk.

These developments demand that we reconceive citizenship for the information age not as a set of rights and responsibilities bestowed by the state (in which case using IT to more efficiently manage the balance sheet of rights and responsibilities is perfectly logical) but as a set of relationships that push and pull us between the poles of making demands and honoring obligations. Technology is not simply a tool used by the state to manage unruly citizens; it is an actor on the shifting terrain of citizenship claims. Each information system is a technology of citizenship.

Over the course of my four years in the YWCA community, I came to believe that the goal of high-tech equity programs should not be to create proficiency or technological skill but rather to produce critical technological citizens who can meaningfully engage and critique the technological present and respond to the citizenship and social justice effects of IT. Rather than teaching specialist knowledge such as software programs and computer functions, activities focused on critical technological citizenship forge links between people's own knowledge of their everyday experience and the "social and political realities" that frame their understandings.[14] Focusing on citizenship rather than on technical proficiency opened up space for women in the YWCA community to learn computing skills to address concrete needs on their own terms, to understand and claim political identities, and to express core values such as justice, recognition, self-determination, and solidarity.

Technology: Understanding Practice and Politics

For the past twenty-five years, the social implications of IT have been a source of almost constant debate among social scientists and theorists. Their analyses have run the gamut from fanatical pronouncements of IT's ability to sweep away social inequalities, reinvigorate government, and create a harmonious global village to strenuous warnings about Big Brother, people becoming robots, and scientists playing God. I have chosen to investigate actually existing technology rather than to track the crest and curve of such overblown claims.[15] We have heard much about the end of work, the twilight of the nation-state, and the triumph of ideas and information over the material, but a critical examination of the role technology can play in fostering progressive social change is as relevant as ever.

Exploring the interaction among technology, citizenship, and social justice requires evaluating what Ursula Franklin calls "real-world technology." Franklin understands technology not just as a set of artifacts but as *practice*, a way things are done. Defining technology as practice "takes machines and devices into account, as well as social structures, command, control, and infrastructures." She writes,

Like democracy, technology is a multifaceted entity. It includes activities as well as a body of knowledge, structures as well as the act of structuring. Our language itself is poorly suited to describe the complexity of technological interactions. The interconnectedness of many of those processes, the fact that they are so complexly interrelated, defies our normal push-me-pull-you, cause-and-consequence metaphors. How does one talk about something that is both fish and water, means as well as end? (Franklin 1999, 35)

For Franklin, attempts to describe the complex interrelations of technology and society are too often collapsed into dualistic ultimatums: Data security or personal privacy? Market mechanisms or human rights? She argues instead for the bridging concept of "adequacy"—of "good enough" understandings, practices, and artifacts. Like Donna Haraway, she insists on the usefulness of provisional, rather than timeless and universal, knowledge.[16] She asks only that these constellations hang together long enough to facilitate collaboration and produce social change.[17] By finding out as much as possible about how women in the YWCA community actually encounter technology, we can begin to understand technology as an everyday practice, and explore its impacts on our relationships and on our notions of power, participation, transparency, and accountability.

How do we reconstruct the house that technology has built? We must start by facing the real world of IT, seeking answers to questions most scholars and policymakers have not yet found the will to ask: What are the *actual* patterns of IT use by women struggling to meet their basic needs? How do they support or contradict the assumptions of public policy? What role does low-wage, contingent labor play in sustaining various high-tech economies and development plans? What political lessons do poor and working-class women learn from their interactions with technologies of citizenship? Starting from the everyday experiences of women in the YWCA community to find empirical answers to these questions can help shape more just policy, better technologies, and more comprehensive solutions to the challenges of high-tech inequity.

Methodology: Studying and Acting Together

In important ways, this book is about method, about how scholars can learn to listen better and to approach our work with a modesty borne of respect for our own limits. It describes a new method for developing high-tech equity programs—popular technology—that starts from the assumption that poor and working-class people already have vast experience with IT and thus come to technology and social justice programs as knowledgeable and asset-bearing rather than deficient or needy. The challenge of popular technology is to turn negative technological experiences into a resource for, rather than a barrier to, learning and engagement. Popular technology takes seriously *all people's* everyday interactions with IT, not just the experiences of the privileged, and then uses these experiences as the starting point for exploring what it means to be a critical citizen in an information age.[18]

Popular technology assumes that collectively produced analysis of structural inequality in our everyday lives provides a source of empowerment, new knowledge, and transformation. This orientation derives from popular education methods, best represented in the United States by the work of the Highlander Research and Education Center. Myles Horton declared that the purpose of founding the Highlander Center[19] in 1932 was to "train community leaders for participation in a democratic society" and to help spread democratic principles to all human relationships in every political, economic, social, and cultural activity (Horton, Kohl, and Kohl 1998, 61–62). Just as for popular education, the fundamental premise that drives popular technology is that people closest to problems have the best information about them and are likely to be most invested in creating smart solutions.

This can be a difficult premise to sustain in the academy. Our training as scholars is centered on mastery, on developing expertise in a specialized subject, and research universities encourage and reward ownership over tightly constrained areas of unique insight. The situation becomes even more complicated when research, education, and activist efforts include IT. As Randy Stoeker has argued, IT is "highly complex, often abstract, and largely the domain of a near-priestly expert class of scientists and network administrators." He writes,

[P]opular education emphasizes the people teaching themselves, rather than being told by outsiders what they should learn and how they should learn it. And that is a particularly challenging leap for those of us in Community Informatics to make . . . few of us can imagine a person who has never sat down in front of a computer being able to work it without us. But that is also what they said about politics, and popular educators have shown them to be wrong there. (Stoeker 2005, n.p.)

Too often in academia, as in most specialized technical fields, we lack the opportunity, motivation, or support to follow where community members outside our universities lead us. When we are able to create the space to do so, our results make significant and innovative contributions to both knowledge and social change. I am not suggesting that every research project should be tackled with participatory methods. But in this case, as I joked with both my YWCA collaborators and my academic colleagues, if participatory action research hadn't existed, we would have had to invent it.

My own understandings of high-tech equity had been so colonized by digital divide theory that I couldn't hear past my own assumptions. Because I was working in a community often tapped for "research subjects," and because the YWCA was an institution committed to economic and racial justice, I had a mandate to share the risks and benefits of the research as equitably as possible. In addition to this ethical obligation, I had empirical and epistemological commitments to producing a better account of the information age, a solid and truthful foundation on which to build a more equitable vision of the world. I spent two years—sometimes daily—in the YWCA community before I asked a single interview question. Had I begun my research with more traditional methods, or from a stance of critical distance and neutrality, I might have gotten this all terribly wrong, seeing only a group of women who were "technology poor" and reproducing the prevailing wisdom.

This book proves the wisdom of digging where you stand and staying with a locally grounded project over the long haul. The groundedness of participatory action research offers a connection to the everyday lives of people that many scholarly accounts of inequality lack.[20] New understanding arose not from choosing an immaculately randomized representative sample but from collaborative analysis. Validity came from a deep connection and passionate engagement within the community—*my* community, my neighbors and friends—not from critical distance and neutrality.

Collaboratively creating the knowledge represented in this book has taken the time, commitment, and effort of dozens of people over the course of a decade. But this kind of work may offer a corrective to high-tech myopia and our dangerously magical thinking about inequality in the information age. Our current understandings cannot account for, and may even obscure, too many critical issues. Over the course of this project, I learned to follow where my neighbors led. They led me to popular technology. I hope their stories will help all of us understand technological citizenship as if poor and working-class women mattered, and in so doing create a new vision of the possibilities for high-tech equity.

3 Trapped in the Digital Divide

Technology is not a destiny but a scene of struggle.
—Andrew Feenberg, *Critical Theory of Technology* (1991, 14)

To understand the analysis offered by women in the YWCA community, and to imagine new possibilities for high-tech equity, we must release our stubborn attachment to the digital divide. The phrase "digital divide" was coined in 1996 by Lloyd Morrisett, a founder of the Children's Television Workshop and president of the Markle Foundation, to describe the chasm that purportedly separates information technology (IT) haves from have-nots in U.S. society.[1] The phrase first captured media and public attention in the late 1990s, when the U.S. National Telecommunications and Information Administration (NTIA) released two reports, *Falling Through the Net II: New Data on the Digital Divide* and *Falling Through the Net III: Defining the Digital Divide*. The reports concluded that, contrary to popular opinion, which held that market forces would eventually lead to universal access, the digital divide had not decreased but increased:

The data reveal that the digital divide—the disparities in access to telephones, personal computers (PCs), and the Internet across certain demographic groups—still exists and, in many cases, has *widened significantly*. The gap for computers and Internet access has generally grown larger by categories of education, income, and race. (NTIA 1999, 2)

The release of the NTIA reports mobilized scholars, politicians, local community-building organizations, and international nongovernmental organizations to inquire whether or not IT could address the needs of the poor, and how exactly to "fit" marginalized people into the information society.[2] The concept has produced much positive scholarly, policy, and activist work, including the community technology center movement and President Clinton's promise to wire every classroom in the nation by the year 2000 and every home by 2007.[3]

Programs dealing with the digital divide and technological opportunity, most of which were underfunded or dismantled during the George W. Bush administration, are beginning to be revived under President Obama.[4] Though the Obama administration has carefully avoided using the phrase "digital divide," expanding broadband access ranks high on the economic agenda.[5] In a December 2008 address, for example, the president-elect promised to renew the nation's information superhighway, extend broadband access to underserved areas, and offer every child a chance to get online. Building Internet infrastructure, he argued, would create millions of jobs and strengthen America's competitiveness throughout the world, leading some technology writers to suggest that "the Internet could prove to be our path to economic salvation" (Karr 2008).

The resilience of the digital divide concept may stem from its resonance with core beliefs in the American political arena. First, the digital divide assumes that technological innovation will automatically lead to social progress, an idea that has had remarkable currency in the Unites States at least since the 1840s (Marx 1993). If technological innovation is synonymous with progress, then the distribution of the products of that innovation becomes a major social justice goal. Second, seeing the problem as a "divide" between technological haves and have-nots keeps the focus on *stuff*, on consumer objects, which fits well in the context of global capitalism. As Darrin Barney (2000) argues, the moral imperative of "access to technology" corresponds to the particular demands of late capitalist economies, specifically the demand for individuals to consume more products produced by high-tech industry (66). Finally, digital divide policy relies on and reinforces the popular idea that there is a self-reproducing "culture of poverty" in the United States that is driven by the individual choices of poor and working-class people themselves rather than being the structural effect of class, racism, and unequal income distribution (O'Connor 2001). Folks trapped in the culture of poverty, digital divide policy posits, simply need to be introduced to a wider world of possibility—perhaps through the Internet—and bootstrapped into the new information age.

Digital divide policy attempts to provide technological tools and develop community organizations that respond to a perceived lack of access or lack of information among supposed technological have-nots.[6] Against their creators' best intentions, however, many digital divide programs actually work to restrict the scope of the high-tech equity agenda because they rely on a deficit orientation that labels neighborhoods "poor" or "underserved" and therefore underestimate the considerable resources, skills, and experiences of these communities. These programs can obscure how powerful institutions such as the criminal justice system, the social service system, and the low-wage workplace operate to structure people's relationship to IT. They also privatize and individualize high-tech equity issues as access issues, limiting opportunities for social mobilization. Most technology policy, firmly planted in the tradition of universal access, ignores nondistributional issues and misrepresents the empirical realities of living in the information age, offering individualized and market solutions to broadly structural problems. The overreliance on the distributive paradigm in digital divide policy and programming is at the heart of our inability to recognize and address some of the most pressing social justice issues of the information age.

Alternative Articulations: Reenvisioning High-Tech Equity

Women in the YWCA community have copious direct experience with information systems and provided extremely astute critiques of the ways that IT is deployed to limit their dignity, freedom, and opportunities. However, despite their ambivalence, most women in the YWCA community showed remarkable perseverance when trying to use IT tools, and even some optimism in describing the possibilities of technological change. When I asked the women I interviewed to finish the sentence "A computer is like . . . ," they responded with "a window," "the future," "a lifesaver." They consistently disproved statements in reports such as *The Ever-Shifting Internet Population*, by the Pew Internet and American Life Project (2003), that claim that "technological pessimism" is a significant barrier to marginalized population's participation in the "information age" (41).

When asked to imagine a socially just information age, women in the YWCA community suggested innovative and incisive new frameworks within which to understand high-tech inequality.[7] For example, when I doodled a picture of the digital divide in my first interview, with Ruth Delgado Guzman (figure 3.1), it visibly upset and frustrated her. Ruth admitted that she did sometimes feel out of the loop, "like a dinosaur," because she doesn't have a laptop, but she was angry that people who lacked access to technology were being seen as needy, deficient people. She explained that people on *both* sides of the putative divide have skills, strengths, and resources to share with each other.

If the digital divide notion was not capturing her experience with IT, I asked, could we describe the problem—and its potential solutions—better? She answered, and I drew what she narrated on top of my original sketch of the digital divide (figure 3.2). Technology, she argued, in the best-case

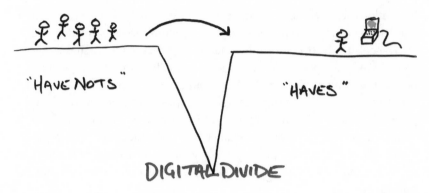

Figure 3.1
My drawing of the digital divide.

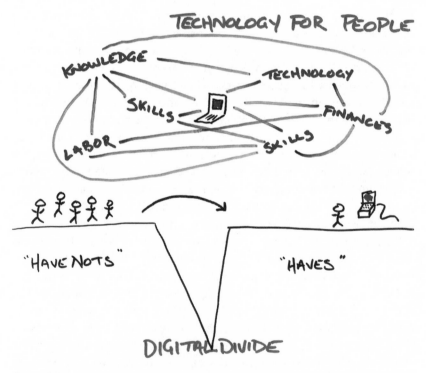

Figure 3.2
A copy of Ruth Delgado Guzman's rearticulation of the digital divide. Reproduced by the author. The original was drawn in the margin of her informed consent form, July 29, 2003.

scenario, should connect people—strengthened by their diverse experiences, across levels of social stratification—in systems of equal barter and exchange. Gesturing to the network drawn in our new model of "technology for people," she said to me, "If you take one message from our conversation to policymakers, it's this. *We don't need to look at the hole. We need to look at the net.*"

This early exchange was an important moment of both empirical and methodological insight: after I interviewed Ruth, I worked "doodling the divide" into many of my interviews, and started carrying acetate overlays for sketching on multiple layers. It proved to be an enormously productive technique for breaking through pat responses to interview questions that simply reproduced magical thinking and popular rhetoric. The exercise provided insights into the critical ambivalence that women in the YWCA community felt about technological change. The sketches that resulted from this process, taken together, illustrate three major critiques of digital divide rhetoric and policy: (1) the characterization of haves and have-nots is overly simplistic; (2) the divide is actually a product of social structure and institutionalized inequalities; and (3) alternative solutions to high-tech equity dilemmas should be developed that leverage technology and diverse local knowledge to build networks based on truth, trust, reciprocity, and reconciliation.

The women I spoke with argued that the categorization of IT users into haves and have-nots is overly simplistic; it does not describe their experiences and obscures structural inequality. For example, Roberta Cousins renamed the haves "technology hoarders" and the have-nots "technology survivors," or "the man" and "the rest of us," respectively (figure 3.3). Other interviewees explained that people in different social and structural positions have access to different kinds of material and intellectual resources—influencing the argument I make in this book about the importance of social location.

While it may be true that folks on the have-not side lack the consumer power that folks on the have side possess, women in the YWCA community insisted that have-nots possess many different kinds of crucial information and skills: community knowledge, knowledge of "the system," double consciousness, and more finely attuned social Geiger counters, as well as social networks, navigation skills, and an ethic of sharing. Therefore, they argued, the have/have-not binary should be reimagined and renamed. Jes Constantine, for example, renamed it the "people divide" (figure 3.4), arguing that the medium was irrelevant and that thoughtful participation, action, and collaboration are the only route to the openness and respect that make communication across difference possible.

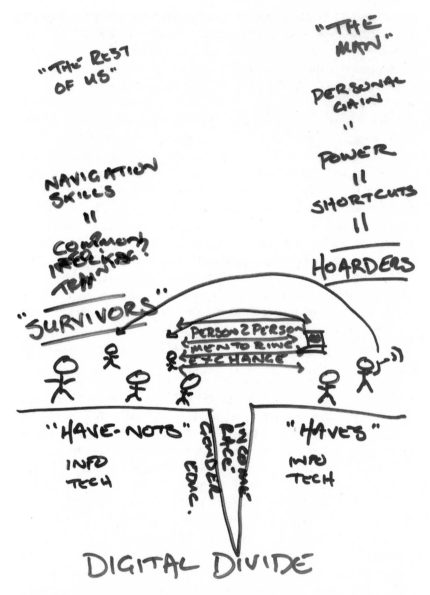

Figure 3.3
Roberta Cousins's articulation of the digital divide. She was speaking, and I was drawing. From our interview, January 19, 2004.

Figure 3.4
Jes Constantine's "people divide." From our interview, January 20, 2004.

Structures and Institutions

Several women specifically pointed out that the digital divide was actually a product of social structure and institutionalized inequality. For example, Jenn Rose renamed the digital divide "social inequality," and Cosandra Jennings renamed it "a crack in the system." Dorothy Allen, Cosandra Jennings, and Roberta Cousins all argued that systemic inequality would not persist if someone were not profiting from it, thereby making space in our discussion for concepts such as oppression, exploitation, and privilege. Jennings pointed out, for example, that both labor and money go from the have-not to the have side to support technological development, and observed that there is a "systemic payoff in [the] disconnect" between the privileged and the exploited. Extraction of resources from the poor to profit the wealthy is represented in Jennings's drawing by the red circle of the system, the money and labor arrows that point from left to right (figure 3.5). This is a sophisticated critique that encapsulates the move from talking about individual choices to talking about systemic power, structures, and institutions.

Dorothy Allen described the historical persistence of structural inequality when she explained that "have-nots come from have-nots" and "haves come from haves" (figure 3.6). She also stressed the role that social, economic, and political privilege play in creating the digital divide, naming social capital, status, and consumption as motivations for the haves to become "information keepers," invested in hoarding information resources and reproducing systems of inequality. This critique is particularly important because it challenges the distributive paradigm at its very core. While distributional frameworks understand injustice as originating in the needs or deficiencies of marginalized populations, Dorothy saw the problem as one rooted in privilege. The people on the have side of the divide are invested in staying on the have side, she, Delgado Guzman, and Jennings all argued, and it is their privilege that must be challenged.

According to the analysis of women in the YWCA community, the problem is not distribution; it is power, privilege, and oppression. Nearly all of the women who sketched the divide with me argued that for this reason, technology alone has little chance of significantly affecting social inequality. More pressing, they argued, are issues of racial prejudice, greed, classism, economic exploitation, basic needs, education, and other social supports—issues that cry out for sustained social activism and major social, structural, and institutional change.

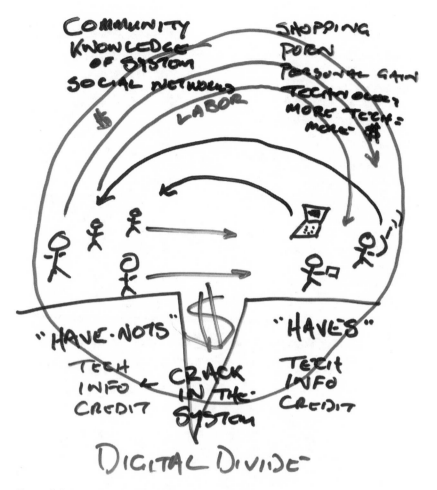

Figure 3.5
Cosandra Jenning's articulation of the digital divide: a crack in the system. From
our interview, January 24, 2004.

Figure 3.6
Dorothy Allen's articulation of the digital divide. The vertical lines at the bottom represent the grass roots. She was speaking, I was drawing. From our interview, February 2, 2004.

Alternative Solutions

The women in the YWCA community offered alternative solutions that leveraged technology and diverse local knowledge to build networks based on truth, trust, reciprocity, and reconciliation. The problems they described, while daunting, are not insurmountable. Jes Constantine, Dorothy Allen, and Roberta Cousins proposed more people-centered than network-centered solutions. Cousins suggested person-to-person mentoring and exchange. Constantine wrote in large capital letters across her drawing, "Technology won't do a single thing unless the people on either end can work together." Allen called for a grassroots social movement that bridges the gap, overcoming blame and ignorance through people's willingness to share their experiences and the reciprocal desire to understand the experiences of others.

Some saw a role technology could play in creating positive social change, seizing on IT's ability to act as an interface across social differences. Jennings argued that IT can be used to educate people on the have side of the divide about the realities of life on the have-not side. Jenn Rose also had faith in the networking potential of IT, using IT as one of four nodes, along with "neutral" space, education and trust building, and media, to create a network of equal exchange across social structures (figure 3.7).

From Distribution to Justice

In the past decade, science and technology policy analysts have endlessly proliferated "divides"—digital, nano, biomedical, MRI, reproductive technology, broadband—when framing high-tech equity problems, and therefore have seen only have and have-not populations, imagined only distributive solutions. This is a grossly oversimplified framework. Instead, following the insights of the women in the YWCA community, we must use *oppression* as the central diagnostic for high-tech equity work. Using oppression as a framework allows us to see complex processes of exploitation, marginalization, powerlessness, cultural imperialism, and violence[8] as central to the inequities of the information age. Identifying these systems of domination provides opportunities for intervention and change. In addition to focusing on getting poor and working-class women into the pipeline, we need to explore how the exploitation of their labor already supports and sustains the information economy. As we struggle for universal participation in accountable, transparent e-governance, we also need to acknowledge how state administration technology currently marginalizes the poor and working-class, contributes to political powerlessness, and increases physical and economic vulnerability.

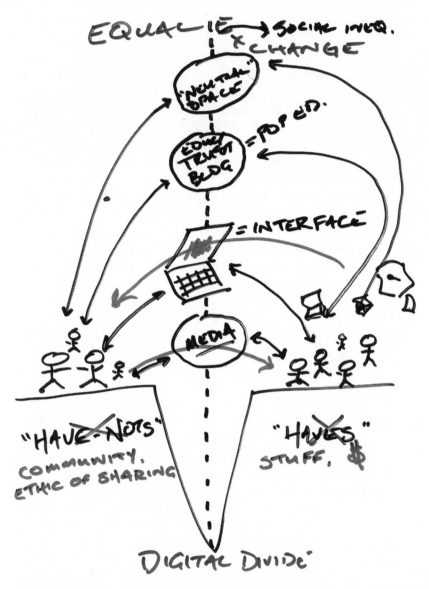

Figure 3.7
Jenn Rose's articulation of one solution to the digital divide: creating nodes of equal
exchange across social stratification. She was speaking, and I was drawing. From our
interview, January 11, 2004.

Box 3.1
WYMSM Member Profile, Cosandra Jennings

"I'm a strong, struggling female."

I'm someone that never gives up, regardless of my whereabouts financially, mentally, physically, emotionally. Finances take a toll on all of it. I'm trying to make ends meet whatever way I can. [When I joined WYMSM,] I was a single mother, with a young child, on social services. I became homeless, and did everything in my means to make extra money, because social services didn't pay me enough to survive on. I had to do what I had to do.

WYMSM gave me more respect for myself, instead of feeling ashamed. It was a big part of my life at that time, to be part of a community with other women, and to hear what they're going through and their struggles. I found out that a lot of women have the same problems. There was a lot going on—my child's father was in prison, I was homeless at the time. So it was a booster for me, it helped me. Even though I was going through a lot, it helped me a lot. I enjoyed it. I definitely miss the group.

I liked the social movement. I felt like I had control over my situation, even though financially, I didn't! But I felt like I had a little more control, knowing that I did something politically. My opinion was heard. And now this book—it feels like, I didn't write the book, but I helped it, I contributed to it. My voice was heard.

I learned a lot about how society works, the government works. I really didn't have knowledge of where things came from—money, politics. I knew a bit about politicians, but it wasn't a big thing for me until after WYMSM. When I vote now, I know what I'm voting for. That's something that is still a big thing in my life, in terms of who I vote for, the community, the social. It's all very important to me.

The thing I most remember about WYMSM was being open and honest about our financials. Not a lot of people are. That was really important to me—to be able to be open about how I lived, about how I had to do things to feed my family and take care of myself. WYMSM taught me that I'm not the only one out there that does them things.

You can feel so alone in this world as a struggling woman. It was a waking moment. Most of us came from poverty, and it was powerful to see these women being honest about the things that they had to do to put food on the table. Hopefully one day it will be recognized, and there will be a change.

My favorite moment was helping get funding for the Women's Resource Directory, to help other women find out how they can get help when they're living in poverty: clothes, children's things, other money that is available to them. That was very interesting to me, and it was very helpful to other people.

Box 3.1
(continued)

We had the group, then we also tried to figure out other ways to help us survive, legally.

Five years from now, I want to be more well off! I want to be stronger as a person, I want to have more communication with the community. I want to not need any services at all from anyplace. I think society should change. People who are very well off need to recognize, and have respect for, people that come from poverty, who are doing the things they are doing to feed their families and to make a living for themselves. Have more respect. I want to tell the reader that, no matter what, if you're struggling, and you're a single mother, that things are going to be hard, but they'll get easier . . . eventually. You have to have some hope.

Based on a conversation over lunch, April 19, 2009.

The three critiques that arose from our exercise of doodling the divide do not fit easily into the framework of the distributive paradigm. In interviews, public forums, and day-to-day personal interactions, women in the YWCA community articulated sophisticated analyses of the relationship between IT and inequity. As is clear in the sketching exercises above, when I talked to women in the YWCA community about developing a high-tech equity agenda, their problem frames (and their solutions) were broadly structural, not distributive. They were concerned with justice, not access.

Framing high-tech equity in terms of oppression rather than distribution makes structural, systemic, and institutional analysis possible. It makes legible the experiences of women in the YWCA community as a crucial part of economic and political restructuring in the wake of neoliberalism, devolution, and "free-market" globalization. Using oppression forces a high-tech equity agenda to deal with social groups rather than individual consumers of technology. Most important, framing high-tech equity issues in terms of oppression allows us to account for differences in group experience grounded in social location.[9] Rather than thinking about access, we need to think about power.

4 Drowning in the Sink-or-Swim Economy

We heard about Rosie the Riveter. And there we were, the YWCA, providing services like childcare and housing for those women who were training for the "high-tech" jobs of their time. And to be frank, we're tired of plugging those holes—the same holes that we were plugging a hundred years ago. . . . We're listening to women. And one of the things they say is that they want more technology opportunities. But they are economically disempowered, and that makes everything else difficult.

—Christine Nealon, Sally Catlin Resource Center director, YWCA of Troy-Cohoes, speaking at the New York State Assembly Roundtable on Women and Technology, April 15, 2003

Women in the YWCA community's call to attend to oppression and power while exploring the relationship between technology and inequality became an antidote to magical thinking and a touchstone for the research and action that followed. Shifting from distributional and individual understandings of high-tech inequity to structural analysis of the information age forced us to reconsider many accepted truths. One of the first assumptions we called into question was the ability of the information economy to lift all boats while delivering regional economic growth and revitalizing cities.

In the early 2000s, Troy underwent a speculative growth spurt when municipal, educational, and business leaders focused on bringing the information economy to the region. Word among homeowners, downtown businesses, commercial developers, and local government officials was that Troy, long challenged by a stagnant economy[1] and declining population base,[2] was turning itself around, injecting new life into its downtown and revitalizing the local economy by attracting out-of-town property buyers, shoppers, businesses, and tourists. These claims were buoyed by a number of factors that converged right around the turn of the millennium. Reverse white flight, the impacts of the events of

September 11, and the skyrocketing cost of living in the New York City metropolitan area combined to usher white, middle-class individuals and families back into the small historic cities that line the Hudson River between Manhattan and Albany.[3] There was a nationwide speculative real estate boom, one that whipped the property markets of small, riverfront cities like Troy into a particularly remarkable frenzy.[4] New economy boosters saw the city as a centerpiece for the region's broader revitalization strategy: high-technology economic development. Troy, home to a well-respected engineering university, enriched by its history of technological innovation and picturesque Hudson River setting, would be the new jewel in the crown of "Tech Valley."

Troy is, in fact, an enchanting city with a rich history and a vibrant population—a wonderful place to live. When I moved here from San Francisco in 1997, I was thrilled by the low rents and happy to see diverse residents, a cross-class mix of African Americans and whites of Italian, Irish, and Polish ancestry, and a growing population of Latinos, Chicanos, and Puerto Ricans. I was fascinated by Troy's incredible legacy of social justice activism. Workers around the time of the Civil War, mostly Irish immigrants, joined together in remarkably successful labor organizing, making Troy the birthplace of a number of innovative and powerful unions, including the Collar Laundry Union, the first sustained female union in the United States.[5] Troy was a center of abolitionism, home to the radical orator and clergyman Reverend Henry Highland Garnet, who published the abolitionist newspapers the *National Watchman* and *The Clarion*, and to family and supporters of Harriet Tubman, who joined with the local antislavery Vigilance Committee in the astonishing liberation of an escaped slave, Charles Nalle, from the custody of U.S. Deputy Marshall John W. Holmes in West Troy in 1860.[6] Troy was a feminist Mecca far back into the nineteenth century, home to the Troy Female Seminary (later Emma Willard School), the first school in the United States to provide young women with an education comparable to that of their male, college-educated peers.

When I first moved to town, the history of Troy seemed to bubble up out of the sidewalks and flow around the cobblestones. When I began to reengage in community activism, around 1999, I felt that I was part of a long tradition. Here in the home of Uncle Sam, I found myself rejecting the cynical hipness of the Bay Area in favor of more all-American fun: homemade ice cream, minor league baseball games, drive-in movies. I even found the Democratic machine politics of the city sort of cute—I would describe Troy to friends back in San Francisco as a teeny tiny Chicago. It was hard to find a coffee on a Sunday morning, and you had to hike over the river or drive up to the sprawl of big box chain stores on Hoosick Street to buy decent produce, but that seemed like a small price to pay to live in such a unique and vibrant city.

Within five years of my arrival in 1997, Troy boosters had firmly affixed their hopes for an urban rebirth to the bandwagon of high-tech economic development, and I began to hear the slogan "Troy is back on track!" all over the region. Like many other cities struggling to stay afloat despite declining industrial and manufacturing sectors, Troy enthusiastically embraced its future as part of a larger initiative termed "Tech Valley,"[7] turning to the information economy in a quest for regional development and municipal stability. State and local governments in the Capital Region—made up of Albany, Rensselaer, and Schenectady counties—provided generous incentives to high-tech companies considering settling in the region.[8] The Tech Valley Chamber Coalition began a yearly Summit in Tech Valley in 2001[9] to highlight the region's technology sector and to promote, market, and brand the region as a high-tech center. Four new tech parks were announced in 2003 alone, and Tech Valley High School, a public magnet school for math, science, and technology, opened in Troy in the fall of 2007.

The region modeled itself openly on Silicon Valley in California and Silicon Gulch in Austin, Texas. Though I was hopeful that Troy could reconstruct its flagging economic base, I was also concerned. I had come to Troy in part because I was fleeing the inequities and the exaggerated rhetoric of the information economy in Silicon Valley. As a person interested in the potential of new information and communications technologies to foster social justice, I was skeptical that the new economy would magically bring material plenty, environmental sustainability, and transparent governance to my new hometown. As an activist concerned with poverty reduction and maintaining the diversity of cities, I was troubled by the way that plans for Troy's rebirth recalled the gentrification I had experienced in the Bay Area. I had lived through the Silicon Valley miracle, and I arrived in Troy fleeing its results. When I started hearing about the plan to create Tech Valley, my hopes for my new home were tinged by worry and apprehension.

I began to question the city's vision for a high-tech boom in earnest after speaking about the information economy with women in the YWCA community. We were very busy at the YWCA in 2002 and 2003, running the Women's Economic Empowerment Series, building a community technology laboratory, and planning the Beat the System game in WYMSM meetings, and I was at the YWCA nearly every day. In conversation, women in the YWCA community told me that things were not getting better for them. In fact, they were getting worse. Affordable housing was harder to find, making it difficult for women to move out of the YW and into the larger Troy community. The waiting list for downtown public housing was eighteen months long—so long, in fact, that the housing authority had stopped accepting applications. Social services were being cut at the local, state, and federal levels. Living-wage jobs, especially unionized work, were harder to come by, as was affordable child care and public transportation. As the furor about Troy's rebirth grew, women in the YWCA were forced to make difficult choices: stay in the YWCA and continue to live in Troy, move back in with local family or friends, or leave the area and try to find affordable housing in other nearby urban centers, like Schenectady or Poughkeepsie.

Public policy in Troy was focused on attracting new, more prosperous residents—at any cost. At the same time that women in the YWCA community were telling me that housing speculation, driven by the new plan for Tech Valley, was raising rents all over the city; officials prioritized attracting young, median-income families and "empty nesters." In a push to increase owner occupancy and decrease rentals, the city targeted housing code violations in Troy's poorest neighborhoods, removing inadequate rental housing stock in the city without creating plans for providing new affordable housing. The city's displacement strategy was made explicit in Troy's 2005–2009 Consolidated Plan, which endorsed reducing public housing in the city by "not replacing units that become obsolete" and suggested that Section 8 voucher holders and public housing residents be advised to relocate outside city limits (City of Troy 2005, 6). The city also counseled neoliberalizing poverty services, asking service providers in Troy to shift their emphasis from "providing institutionalized supports and towards services that promote a transition to self-sufficiency," focused mostly on job training, life-skills counseling, and day care (ibid., 7). In the midst of massive public investments in the high-tech economy in the Capital Region, low-income residents were being asked to pull themselves up by their bootstraps or get out of the way.

I was dismayed by these paradoxical policy imperatives. On the one hand, policymakers and municipal boosters argued that technology and high-tech economic growth are a route to empowerment of the poor and disenfranchised and a silver bullet solution to regional economic decline. Community technology centers proliferated in low-income neighborhoods, high-tech companies competed for huge incentives to move to the region, infrastructure was created or rerouted to facilitate the coming of Tech Valley. On the other hand, policymakers and municipal boosters created policy that would clearly result in the displacement of Troy's most vulnerable people. Public housing was allowed to deteriorate, rents sky-rocketed, and social supports were harder to get and keep. From the stand-point of the YWCA, these policies seemed unmoored from reality. It was Silicon Valley all over again: the information economy was going to lift some boats, but it was a comfortable ride only if you could ignore the screams of the people drowning all around you.

In this chapter, I use an intersectional approach to explore claims made by high-tech economic development boosters that the information economy is a force for equity. Intersectional feminist analysis of the infor-mation economy demands that we ask more difficult questions than "How do we fill the pipeline?" It should question taken-for-granted assumptions, center itself in explorations of power, oppression, and domination, and uncover features of social life that have been made invisible by dominant economic models (Barker and Feiner 2004, 9). It demands that we ask, "Who pays the price for regional economic development?," "Which workers do we see when we talk about the high-tech economy?," and "How is the information economy playing out for marginalized people in the real world?"

Volatile Continuity in the Information Economy

I was having conversations about the difficulties women in the YWCA community were having finding affordable housing and adequate social services in Troy during the "jobless recovery" from the recession that fol-lowed the dot-com crash. Though the reality of the crash had not soured the optimism of Tech Valley advocates, women in the YWCA community were feeling its effects keenly. Local rhetoric proclaimed Troy on the verge of a comeback, extolling mass purchases of downtown buildings by local and out-of-town developers,[10] celebrating the creation of antiques districts and waterfront condos, and waxing optimistic about Troy neighborhoods returning to life. But women in the YWCA community were feeling increas-ingly displaced and vulnerable.[11]

In June 2003, Harris Collingwood asked a question that resonated with women in the YWCA community: "If the recession isn't so deep, how come you're feeling so bad?" In his *New York Times Magazine* article, "The Sink-or-Swim Economy," Collingwood argued that the GDP had grown slug-gishly but steadily throughout 2002 and the first quarter of 2003. And yet,

Box 4.1

WYMSM Member Profile, Chitsunge Mapondera

"You do things that you love, but you do them in such a way that they can also have a positive impact on other people."

I got involved with the YWCA because I was very eager to do something beyond the walls of Rensselaer Polytechnic Institute, where I was an electrical engineering student. What's always made sense to me is to be self-directing, to create an opportunity for myself and go after it. Some of the most important lessons do not come to you in a formal setting where you sit down and say, "OK, I'm ready to learn now." I also felt like I had something to give, something to offer, even if it was just my attitude and how I represented myself. I could present an image or idea of what a good man, a decent man, is supposed to be.

I come from Africa, from Zimbabwe. In the town where I grew up, you had very severe inequity. Mainstream society was very well-to-do, but right next door was an impoverished community. Seven of the ten houses on your street would have electric gates and chauffeur-driven Mercedes-Benzes, but people at the end of the street lived under a tree. It was right under your nose, it surrounded you. So coming from Africa, I've never been confused about poverty. Then, I was in Troy, in America! And the image I had of America was very different from what I'm seeing at the YWCA, which is the same kind of impoverishment, the same kind of complexity and difficult situations, though they manifest differently.

Life in the State of Poverty was one of the most important things WYMSM did. The game inherently created relationships: you had different roles, different families, different characters. On top of that you have actual people behind all those characters, so now you're also establishing relationships between people. When we debriefed at the end, women [from the YW community] were in control. They were in charge. There was a connection that was happening, a very powerful energy. Their stories and life experiences, their daily lives, affected how the game played out. They were able to be the teachers.

What also stands out for me is all the people who made the YWCA work. That was a genuine group of people, very honest. A lot of people living at the YW were coming from very complex, complicated situations. In their own way, the staff made life simple. And at the same time, the whole experience was fun! It remains one of the fondest times of memory in my life, as far as the quality of people. Even though me and Jes spent many a day looking at each other with evil eyes as we went through and put that computer network through all five floors of that building—and the basement!—I don't think I

Box 4.1
(continued)

ever left the YW in a bad mood, with a bad attitude. There wasn't a day that you'd leave and carry any bad feeling with you.

The YWCA brought out my better qualities—it was a confidence-building experience, a coming of age. If I think of myself as a future deal-maker, that was my first table. Ten years from now, I'll be moving between the U.S. and Africa. I'll be managing private equity and venture capital funds in Southern Africa, and will be investing in education, information and communications companies and technologies, infrastructure, and vital services and resources that people need—farming, food production, transportation, building and housing. I guess you'd call it alternative investments. With WYMSM, I developed a desire to do things that have a net-positive impact on society. You're never truly satisfied doing what you do just because you love it. At the end of the day, you do things that you love, but you do them in such a way that they can also have a positive impact on other people.

Based on a phone conversation that took place June 1, 2009.

he wrote, the economy shrank by half a million jobs in the first four months of 2003, and most wages stagnated or declined. These contradictory indicators—modest but enduring GDP growth and continuing declines in employment—are not entirely remarkable, and are often written off by economists as statistical noise. But in the new sink-or-swim economy, Collingwood argued, these opposing trends were becoming permanent. "They call it 'idiosyncratic volatility,'" he wrote, "and it is the signature of our economic age" (Collingwood 2003, 44). The increasingly vicious swings of the American economy in recent years have only underscored Collingwood's insights into the volatility and risk of the new economy, as their impacts penetrate deeper into the middle class.

In the popular press, accounts of the information economy posit that increased instability and volatility can offer more horizontal forms of power, free workers to retool their skills and renegotiate their work arrangements, and sweep away old forms of inequity.[12] The combination of new IT and leaner, neoliberal governance, optimists argue, results in rapidly increasing wealth and flatter hierarchies, although these claims have been somewhat muted in recent years.[13] The most popular of these narratives, penned by business writers, futurists, and management gurus, often make it to the bestseller lists, suggesting that they tap into widely held hopes and beliefs about the power of IT and the new economy to dismantle out-of-date institutions, decentralize power, and create broad-based equity.[14]

For example, Kevin Kelly, executive editor of *Wired* magazine, argues in his 1998 book, *New Rules for the New Economy*, that the network economy is based on the principles of flux. He writes, "Change, even in its shocking forms, is rapid difference. Flux, on the other hand, is more like the Hindu god Shiva, a creative force of destruction and genesis. Flux topples the incumbent and creates a platform for more innovation and birth" (10). Thomas L. Friedman makes a similar argument in his 2005 bestseller, *The World Is Flat*, explaining that digital connectivity produced rapid changes in the last two decades—including the fall of the Berlin Wall, the invention of the Netscape Web browser, and employment practices such as outsourcing and off-shoring—that act as flattening and leveling forces, creating broad-based equity across the globe. He writes, "[F]lattening forces are empowering more and more individuals today to reach farther, faster, deeper, and cheaper than ever before, and this is equalizing power—and equalizing opportunity, by giving so many more people the tools and ability to connect, compete, and collaborate" (x).

Volatility, risk, and flexibility are indeed signature features of the information economy. But they are not nearly as idiosyncratic as these writers assume. The new economy, digital technology, and laissez-faire governance do not, in fact, annihilate existing configurations of power, sweeping away the wreckage of the old, leaving a playing field that is clean, smooth, and level. Rather than magically clearing away old structures of inequality in a tide of creative destruction, the information economy is the product of two equally significant forces: increasing macroeconomic instability *and* comparatively stable, preexisting topographies of social and economic inequality. Rather than being unpredictable or idiosyncratic, the vulnerabilities and risk produced by the information economy display a great deal of continuity with what came before.[15] I call this phenomenon *volatile continuity*. It is endemic to the information age, and it produces new configurations of inequality that are historically, contextually, and geographically specific.[16]

Volatile continuity creates a dual economy marked by an increase in professional, high-wage jobs that require a great deal of educational and social capital investment and a parallel increase in the service and caregiving industries, where wages are stagnant or dropping and labor protections are rare. Where high-tech development succeeds, an influx of executives, managers, and engineers rapidly inflates housing markets, driving up the cost of living and creating a secondary economy made up of the poor and working-class, who take care of the personal, child-care, elder-care, health, and retail needs of the so-called "creative class."[17] After a brief boom in construction and light manufacturing industries, electronics manufacturing—like chip fabrication, the industry most eagerly sought by Tech Valley—often moves to another region or offshore, leaving behind only those workers who cannot (or will not) be deterritorialized: on the one hand, executives, engineers, and creative professionals, and on the other, janitors, child care workers, and restaurant employees.[18] If broad-based educational and economic equity does not already exist when the high-tech economy arrives in a given area or if it is not consciously protected through aggressive equity policies, local workers disadvantaged by legacies of race, class, or gender discrimination are poorly placed to take advantage of the opportunities offered by the high-tech economy and are likely to take on more than their fair share of its burdens. The ability of workers and other citizens to profit from the increased volatility of the information economy is, thus, closely tied to their social location.

Topographies of Inequality in Troy

The volatility of the information economy is not idiosyncratic because the ruptures and shocks of increasingly rapid change arrive in *actual places*. These places have history, contexts, legacies of past systems of domination, and institutions shaped by continuing oppression and inequity. In Troy, the volatility of the information economy arrived on a landscape already deeply inscribed by racial discrimination and economic stratification. According to the 2000 census, there were striking gender- and race-based inequalities in educational attainment, employment status, earnings, and poverty in the Capital Region and the city of Troy when Tech Valley boosters began extolling Troy's rebirth. More strikingly, according to the 2006–2008 American Community Survey 3-Year Estimates, many of these forms of inequality grew rather than shrinking as the information economy took hold in the Capital Region.

Educational Attainment
Jobs in the top tier of the dual economy are dependent on a highly educated workforce. But educational attainment in the Capital Region and the city of Troy is highly mediated by gender and race. According to American Community Survey 2006–2008 estimates, despite significant improvements in the last decade, Black or African American men over the age of twenty-five still earn 16 percent fewer bachelor's and advanced degrees

than white men in Troy. Black or African American women earn 32 percent fewer degrees than white women. Latinos are actually losing ground educationally—according to the 2006–2008 estimates, men of Hispanic or Latino descent in Troy are earning fewer degrees than they did in 2000, and they are earning 26 percent fewer degrees than white men. Despite a significant growth in degree attainment in the last decade, Latinas still earn 48 percent fewer degrees than white women.

There are also significant disparities in education by gender within racial groups. Though white women have nearly caught up to white men in degree achievement—earning only 2.4 percent fewer degrees in 2006–2008—women of Hispanic or Latina descent still earn 30 percent fewer degrees than Hispanic or Latino men. Black or African American women are actually losing ground in comparison to men in their own racial category: they earned 11 percent fewer degrees than Black or African American men in 2006–2008, compared to 8 percent fewer degrees in 2000. More striking still are differences in educational attainment among women by race: white women in Troy earn one-third more degrees than Black or African American women and nearly twice as many as Latinas.

Employment Status and Earnings
Employment status and earnings provide a rough picture of how many jobs are available in the new economy, and whether wages are growing, shrinking or stagnating. In 2006–2008, unemployment was 75 percent higher in Troy than in the Capital Region, and 44 percent higher than the national average. African Americans in the labor market in Troy faced a 14.5 percent unemployment rate, nearly twice the rate of white workers. Black men faced an astronomical unemployment rate of 16.9 percent in Troy, two and a half times higher than white men and 30 percent higher than the national level. Though city-level data for male Hispanic or Latino workers are unavailable, they are unemployed 60 percent more often than white men (8.4 percent versus 5.1 percent) at the regional level, and women of Hispanic or Latina descent are unemployed more than three times as often as white women (12.7 percent versus 4.1 percent). Bucking the trends at the national and regional level, unemployment in Troy is actually worse for white and Hispanic or Latina women than for men in their own racial category. However, Black or African American women face 28 percent less unemployment than Black or African American men.

Earnings inequality between the sexes is still a fact of life in every racial group, with the wages of women who work full-time, year-round standing at between 76 percent and 90 percent of men's in Troy. The racial gap

Table 4.1
Educational attainment by race and gender, 2000 and 2006–2008

Population over 25 who earned a bachelor's degree or above	United States	Albany-Schenectady-Troy MSA	City of Troy
White men, 2000	29.3% (n = 18,437,073)	30.4% (n = 75,739)	19.9% (n = 2,296)
White men, 2006–2008	31.8% (n = 21,218,917)	33.7% (n = 80,690)	21.2% (n = 2,255)
% Change, 2000–2008	*+8.5%*	*+10.9%*	*+6.5%*
White women, 2000	24.8% (n = 17,323,549)	26.6% (n = 75,739)	18.1% (n = 2,457)
White women, 2006–2008	29.1% (n = 20,902,057)	32.4% (n = 85,703)	20.7% (n = 2,499)
% Change, 2000–2008	*+17.3%*	*+21.8%*	*+14.4%*
Black or African American men, 2000	13.1% (n = 1,176,692)	16.3% (n = 2,203)	9.9% (n = 136)
Black or African-American men, 2006–2008	15.7% (n = 1,580,496)	19.7% (n = 2,694)	16.8% (n = 231)
% Change, 2000–2008	*+19.8%*	*+20.9%*	*+69.7%*
Black or African American women, 2000	15.2% (n = 1,654,577)	14.8% (n = 2,147)	10.7% (n = 162)
Black or African-American women, 2006–2008	18.5% (n = 2,239,693)	15.7% (n = 2,563)	14.0% (n = 200)
% Change, 2000–2008	*+21.7%*	*+6.1%*	*+30.8%*
Hispanic men or Latinos, 2000	10.2% (n = 933,014)	24.0% (n = 1,281)	16.6% (n = 65)
Hispanic men or Latinos, 2006–2008	11.8% (n = 1,508,345)	23.4% (n = 1,517)	15.6% (n = 70)
% Change, 2000–2008	*+15.7%*	*–2.5%*	*–6.0%*
Hispanic women or Latinas, 2000	10.7% (n = 975,025)	23.3% (n = 1,347)	8.5% (n = 34)
Hispanic women or Latinas, 2006–2008	13.5% (n = 1,609,671)	21.8% (n = 1,493)	11.0% (n = 78)
% Change, 2000–2008	*+26.2%*	*–6.4%*	*+29.4%*

Sources: U.S. Census Bureau (2000), tables P37, P148B, P148H, and P148I; U.S. Census Bureau (2008), tables C15002, C15002B, C15002H, and C15002I.

Table 4.2

Unemployment by gender and race, 2000 and 2006–2008

Civilian unemployment for persons 16 and over	United States	Albany-Schenectady-Troy MSA	City of Troy
White men, 2000	4.4% (*n* = 2,367,979)	5.6% (*n* = 12,038)	10.7% (*n* = 1,385)
White men, 2006–2008	5.3% (*n* = 56,766,714)	5.1% (*n* = 81,922)	6.6% (*n* = 4,730)
% Change, 2000–2008	*+20.5%*	*–8.9%*	*–38.3%*
White women, 2000	4.2% (*n* = 1,975,993)	4.3% (*n* = 8,446)	7.3% (*n* = 714)
White women, 2006–2008	5.0% (*n* = 34,371,792)	4.1% (*n* = 117,641)	8.2% (*n* = 5,859)
% Change, 2000–2008	*+19.0%*	*–4.7%*	*+12.3%*
Black or African American men, 2000	12.0% (*n* = 835,490)	15.6% (*n* = 1,735)	12.8% (*n* = 212)
Black or African American men, 2006–2008	12.9% (*n* = 4,701,001)	11.1% (*n* = 7,145)	16.9% (*n* = 624)
% Change, 2000–2008	*+7.5%*	*–28.8%*	*+32.0%*
Black or African-American women, 2000	10.8% (*n* = 862,927)	12.1% (*n* = 1,423)	15.1% (*n* = 181)
Black or African American women, 2006–2008	11.1% (*n* = 5,514,815)	10.0% (*n* = 7,774)	12.2% (*n* = 467)
% Change, 2000–2008	*+2.8%*	*–17.4%*	*–19.2%*
Hispanic men or Latinos, 2000	8.2% (*n* = 708,212)	12.6% (*n* = 630)	15.4% (*n* = 83)
Hispanic men or Latinos, 2006–2008	6.5% (*n* = 3,575,649)	8.4% (*n* = 2,709)	[ND]
% Change, 2000–2008	*–20.7%*	*–33.3%*	—
Hispanic women or Latinas, 2000	10.6% (*n* = 663,629)	12.2% (*n* = 600)	20.8% (*n* = 74)
Hispanic women or Latinas, 2006–2008	8.6% (*n* = 6,333,374)	12.7% (*n* = 3,541)	[ND]
% Change, 2000–2008	*–18.9%*	*+4.1%*	—

Sources: U.S. Census Bureau (2000), tables P43, P150B, P150H, and P150I; U.S. Census Bureau (2008), tables C23001, C23002B, C23002H, and C23002I.

between women's wages is also significant. Black or African American women make 24 percent less and Latinas 51 percent less than white women for full-time, year-round work in Troy.

Poverty

It is clear that women in Troy, particularly women of color, are not yet seeing the rising tide lift their boats. Despite significant strides in educational attainment, women of color face astronomical rates of unemployment and lower wages in comparison to white men and women. It is not surprising, then, that women of color in Troy face calamitous levels of poverty. While the 17 percent poverty rate for white women in Troy is distressing, more troubling still is the fact that in 2006–2008, 29 percent of Black or African American women in Troy—and more than *half* of all Latinas—lived in poverty.

Volatile continuity produces new configurations of inequality that extend and shift past legacies of oppression and discrimination. Inequities are generated or deepened even in those regions held up as models of new economy success, belying high-tech boosters' promises of flatter hierarchies and more level playing fields. For example, in *Work in the New Economy*, Chris Benner's 2002 case study of shifting labor markets in Silicon Valley, he argues that economic volatility is not primarily a leveling or flattening force; rather, flexibility has both positive and negative impacts on workers. Flexibility in Silicon Valley labor markets was a key part of the region's economic success and led to good outcomes for some categories of workers, particularly those with educational advantages.

For other categories of workers, however, increased flexibility, especially in the employment relationship, means increased vulnerability. This vulnerability is manifested in increased outsourcing, temporary and contingent labor arrangements; high levels of "job churn," turnover, and mobility; and high levels of skills obsolescence owing to the rapid pace of technological, industry, and market change (Benner 2002, 39–48). These characteristics create a labor market with significant negative effects for workers already made vulnerable by their position in the social hierarchy and the division of labor, particularly those workers marginalized by gender, class, and race. "Flexible labor markets are risky labor markets," writes Benner, "and many workers suffer from longer periods of unemployment, difficulty acquiring training, and greatly reduced bargaining power" (205).

Flexible labor markets are particularly tricky for workers in Troy. Suffering poverty, unemployment, and educational shortcomings at rates much higher than elsewhere in the region or the country, Troy's poor and working

Table 4.3

Women's median earnings as a percentage of men's, 1999 and 2006–2008

Women's median earnings as a % of men's, for population 16+, working full-time, year-round	United States	Albany-Schenectady-Troy MSA	Troy
Entire population, in 1999 (in 2000 dollars)	73.4% M: $38,467 W: $28,621	74.4% M: $38,467 W: $28,621	84.4% M: $30,495 W: $25,724
Entire population, in 2006–2008 (in 2000 dollars)	77.6% M: $36,794 W: $28,568	77.5% M: $40,707 W: $31,554	83.6% M: $33,593 W: $28,081
% Change, 2000–2008	*+5.7%*	*+4.2%*	*–0.9%*
White alone, not Hispanic or Latino, in 1999 (in 2000 dollars)	70.4% M: $40,160 W: $28,265	74.0% M: $39,313 W: $29,094	84.1% M: $31,029 W: $26,087
White alone, not Hispanic or Latino, in 2006–2008 (in 2000 dollars)	77.6% M: $41,410 W: $30,331	78.1% M: $41,635 W: $32,500	89.0% M: $34,681 W: $30,863
% Change, 2000–2008	*+10.2%*	*+5.5%*	*+5.8%*
Black or African American only, in 1999 (in 2000 dollars)	85.3% M: $30,000 W: $25,589	91.2% M: $27,318 W: $24,914	87.2% M: $27,756 W: $24,200
Black or African American only, in 2006–2008 (in 2000 dollars)	87.3% M: $29,473 W: $25,721	84.4% M: $29,335 W: $24,758	90.4% M: $29,194 W: $26,386
% Change, 2000–2008	*+2.3%*	*–7.5%*	*+3.7%*
Hispanic or Latino only, 1999 (in 2000 dollars)	85.2% M: $25,400 W: $21,634	87.9% M: $28,410 W: $24,984	75.3% M: $22,067 W: $16,615
Hispanic or Latino only, 2006–2008 (in 2000 dollars)	87.0% M: $24,520 W: $21,323	88.6% M: $27,476 W: $24,348	76.3% M: $19,969 W: $15,237
% Change, 2000–2008	*+2.1%*	*+0.8%*	*+1.3%*

Sources: U.S. Census Bureau (2000), tables PCT74, PCT74B, PCT74H, and PCT74I; U.S. Census Bureau (2008), tables B20017, B20017B, B20017H, and B20017I.

Table 4.4
Poverty status by gender and race, 2000 and 2008

Persons living in poverty, of those for whom poverty status is determined	United States	Albany-Schenectady-Troy MSA	City of Troy
White men, 2000	7.1% (*n* = 6,583,265)	6.3% (*n* = 22,886)	14.0% (*n* = 2,420)
White men, 2006–2008	8.1% (*n* = 7,642,154)	6.9% (*n* = 23,527)	14.3% (*n* = 2,348)
% Change, 2000–2008	*+13.4%*	*+8.6%*	*+2.8%*
White women, 2000	9.1% (*n* = 8,830,854)	8.2% (*n* = 31,510)	15.4% (*n* = 2,873)
White women, 2006–2008	10.3% (*n* = 10,123,984)	8.4% (*n* = 30,396)	17.2% (*n* = 2,777)
% Change, 2000–2008	*+12.8%*	*+3.1%*	*+11.3%*
Black/African American men, 2000	22.8% (*n* = 3,446,854)	28.5% (*n* = 6,653)	31.8% (*n* = 833)
Black/African American men, 2006–2008	22.4% (*n* = 3,703,201)	26.5% (*n* = 6,837)	38.2% (*n* = 967)
% Change, 2000–2008	*−1.7%*	*−6.9%*	*+19.8%*
Black/African American women, 2000	24.4% (*n* = 4,181,391)	32.0% (*n* = 8,045)	33.1% (*n* = 878)
Black/African American women, 2006–2008	26.6% (*n* = 5,045,041)	32.1% (*n* = 9,022)	28.9% (*n* = 761)
% Change, 2000–2008	*+8.9%*	*+0.5%*	*−13.0%*
Hispanic/Latino men, 2000	21.1% (*n* = 3,698,911)	25.4% (*n* = 2,740)	35.6% (*n* = 299)
Hispanic/Latino men, 2006–2008	19.0% (*n* = 4,331,387)	22.8% (*n* = 2,826)	58.2% (*n* = 734)
% Change, 2000–2008	*−10.1%*	*−5.8%*	*+63.5%*
Hispanic/Latina women, 2000	24.2% (*n* = 4,098,963)	29.8% (*n* = 3,403)	49.1% (*n* = 485)
Hispanic/Latinas women, 2006–2008	23.6% (*n* = 5,110,611)	30.4% (*n* = 3,933)	55.5% (*n* = 855)
% Change, 2000–2008	*−2.5%*	*+2.2%*	*+13.0%*

Sources: U.S. Census Bureau (2000), table numbers PCT49, PCT75B, PCT75H, and PCT75I; U.S. Census Bureau (2008), table numbers C17001, C17001B, C17001H, and C17001I.

class earn lower wages while trying to manage the higher cost of living and more insecure working conditions brought to the region by the information economy. Further, the plight of workers in Troy is often hidden by geographically aggregated statistics, which can obscure severe localized inequalities resulting from regional economic development schemes. It is not difficult to predict that, barring the creation of an aggressive equity policy, the increased instability of the economy will have disastrous effects for already marginalized workers. Lacking advanced education, unemployed, underpaid, and impoverished, they are unlikely to benefit from the massive public expenditures undertaken to attract high-tech employment to the region.

Has the Information Economy Arrived in the Capital Region?

In the past ten years, New York state legislators have offered incentives to high-tech industry so large that they have had to invent a new term for them—"mega-incentives." These incentives have been a qualified success in bringing the information economy to the Capital Region. Several major projects and collaborations have been launched and billions of dollars, both public and private, have been invested in the high-tech future of the region. Local universities, such as Rensselaer Polytechnic Institute and the University at Albany, SUNY, have partnered with global corporations, including Applied Materials, Advanced Micro Devices (AMD), General Electric, IBM, and Tokyo Electric, to create manufacturing facilities for computer chips and state-of-the-art health care equipment, and research institutes for nanotechnology, electron-beam lithography, and semiconductors. AMD, a microprocessor company based in Sunnyvale, California, and the Mubadala Development Company, AMD's Abu Dhabi–based financial backers, plan to build a major chip fabrication plant in nearby Malta. The State of New York has dedicated more than $10.5 billion in cash incentives and tax breaks since 2000 to attract high-tech companies to the area, or to keep high-tech companies like General Electric and IBM from moving their research centers and chip fabrication plants elsewhere. But what are we investing in when we invest in high-tech?

At the regional level, high-tech boosterism brought more jobs in the information economy sectors in the last decade, but in complicated ways. Jobs in the NAICS category of information—which includes the publishing, software, motion picture, broadcasting, radio, cable, Internet service provider, and telecommunications industries—held steady between 2001 and 2003 (at just over 11,500 jobs) before dropping precipitously over the next five years (to 9,247 jobs), producing a 19.7 percent decline in information category jobs between 2001 and 2008. Other hallmarks of the information economy, such as jobs in the NAICS categories of financial activities—which include financial, insurance, and real estate industries (FIRE)—gained 1,370 jobs since 2001 (just over 5 percent growth), and professional and business services—which include technical and scientific services—gained 4,288 jobs since 2001 (8.6 percent growth). Overall, these three categories of employment have added just over 3,300 jobs to the Capital Region in the last eight years. The chief impact of the information economy in the region has actually been a rapid growth in the service industries: education and health services grew by 7,168 jobs between 2001 and 2008, a 10.9 percent growth; leisure and hospitality grew by 4,282 jobs, a 14.7 percent growth; and "other services" (a category that includes maintenance, repair, foundation, and social advocacy professions) grew by 874 jobs, a 5.8 percent growth. Manufacturing, traditionally a category that provides unionized jobs at livable wages, saw rapid employment contraction between 2000 and 2008, losing 4,899 jobs, or an 18 percent loss. Overall, while we are seeing the two-tiered growth indicative of the information economy, the service sector is growing much more quickly than the knowledge-based sectors, gaining four jobs for every one created in information, FIRE, and professional and business services.

Wages in the top-tier information economy occupations grew significantly (from 6.8 to 8.7 percent) between 2001 and 2008. Wages in the lower-tier service occupations also rose, but more modestly: 6.9 percent for education and health, 5.4 percent for leisure and hospitality, and 5.4 percent for other services. They will likely rise still more as staggered minimum-wage raises, instituted on the federal level, take effect. But wages in the service industries are still incredibly low in comparison to jobs in the goods-producing and top-tier information economy occupations. For example, the information classification's average weekly wage of $937 is 55 percent more than that for education and health services ($604), twice as high as that for other services ($462), and three and a half times the weekly wage of leisure and hospitality ($262).

The number of jobs in goods-producing occupations, which traditionally offered higher wages and stable employment for those without college degrees, is dropping rapidly in the Capital Region. More highly paid, education-dependent, top-tier information economy occupations are behaving erratically. Wage growth in these top-tier job categories is steady and significant, and wages in these categories approach or achieve living-wage standards for the area. However, the most rapid job growth in the

Table 4.5
Change in Capital Region industrial composition and wages, 2001–2008

NAICS industry classification	No. of jobs gained or lost between 2001 and 2008	Employment growth between 2001 and 2008	Average weekly wage, 2008 (in 2000 dollars)	Wage growth between 2001 and 2008
Goods-Producing Information Economy Occupations				
Manufacturing	−4899	−18.0%	$955	+5.4%
Construction	+123	+0.7%	$813	+7.1%
Top-Tier Information Economy Occupations				
Information	−2271	−19.7%	$937	+8.7%
Financial activities	+1370	+5.4%	$880	+6.8%
Professional and business services	+4288	+8.6%	$864	+8.3%
Bottom-Tier Information Economy (Service) Occupations				
Education and health services	+7168	+10.9%	$604	+6.9%
Other services	+874	+5.8%	$462	+2.4%
Leisure and hospitality	+4282	+14.7%	$262	+5.4%

Source: Bureau of Labor Statistics, Quarterly Census of Employment and Wages for the Albany-Schenectady-Troy MSA (U.S. Department of Labor 2008). The tripartite division of the information economy into productive, virtual and reproductive sector was inspired by Peterson (2003).

Capital Region over the last decade has been in the bottom tier of the information economy, in the service sector, which offers low wages and insecure employment relationships.[19]

Does the Rising Tide Lift All Boats?

Though growth in high-tech sectors in the Capital Region has been sporadic, changes in industrial composition, employment patterns, and income polarization *do* point strongly to a structural shift toward flexible, volatile employment. This is a key indicator that, regardless of whether or not we are attracting top-tier high-tech jobs to the region, the information economy has arrived. Has it fulfilled Friedman's promises? Is inequality leveling, are hierarchies flattening? Structural shifts in the Capital Region economy have had mixed impacts on educational, employment, earnings, and poverty inequalities. Table 4.6 shows relational rates of educational attainment, unemployment, median earnings, and poverty levels between men and women within racial categories and among women by race in order to assess changes in inequality.

For example, for educational inequality, I calculate women's rates of educational attainment as a percentage of the rate of educational attainment of men in their own racial group (the first three columns), and rates of educational attainment of Black or African American and Hispanic or Latina women as a percentage of the rate of educational attainment of white women (last two columns) at the national, regional, and city level. Then I provide the same rates calculated using the 2006–2008 American Community Survey averages. Finally, I compare the two to judge whether inequality increased or decreased between 2000 and 2008. I follow the same procedure for unemployment, median earnings, and poverty. The table shows that, despite some important gains, the equalizing and leveling of inequality promised by high-tech boosters has not materialized. In fact, many forms of inequality increased significantly in the region as compared to national levels in the last decade.

Change in Educational Inequality

At the regional level, white women have made great strides toward educational equity with men. But educational inequality between Black or African American women and men and Latinas or Hispanic women and Latinos or Hispanic men is growing. In addition, educational inequality among women by race is increasing rapidly: the gap between Black or African American and white women's educational achievement grew 13 percent between 2000 and 2008, and the gap between Latina or Hispanic women and white women's educational achievement grew by 23 percent. In Troy, the educational gap between white women and men is shrinking. But Black or African American women are losing ground compared to Black or African American men: in 2000, they earned 8 percent more degrees; in 2008, they earned 15 percent fewer. Latinas and Hispanic women earned within 30 percent of the number of degrees of Latinos or Hispanic men. In Troy women of color are making strides to overcome gaps in educational achievement in comparison to white women, though Black or African American women still earn less than three-quarters the number of degrees as white women. Latinas or Hispanic women earn only half.

Table 4.6

Changes in inequality, 2000–2008

Changes in Inequality, 2000–2008		White women as % of white men	Blk/AA women as % of Blk/AA men	Lat/Hisp women as % of Lat/Hisp men	Blk/AA women as % of white women	Lat/Hisp as % of white women
Education	United States	Lower	Lower	Lower	Lower	Lower
	2000	84.6%	116.0%	104.9%	61.3%	43.1%
	2008	91.5%	117.8%	114.4%	63.6%	46.4%
	% Chg.	−8.1%	−1.6%	−9.1%	−3.7%	−7.5%
	MSA	Lower	Higher	Higher	Higher	Higher
	2000	87.5%	90.8%	97.1%	55.6%	87.6%
	2008	96.1%	79.7%	93.2%	48.5%	67.3%
	% Chg.	−9.9%	12.2%	4.0%	12.9%	23.2%
	Troy	Lower	Higher	Lower	Lower	Lower
	2000	91.0%	108.1%	51.2%	59.1%	47.0%
	2008	97.6%	83.3%	70.5%	67.6%	53.1%
	% Chg.	−7.4%	22.9%	−37.7%	−14.4%	−13.2%
Unemployment	United States	Higher	Higher	Higher	Lower	Lower
	2000	95.5%	90.0%	129.3%	257.1%	252.4%
	2008	94.3%	86.0%	132.3%	222.0%	172.0%
	% Chg	1.2%	4.4%	−2.4%	13.7%	31.8%
	MSA	Lower	Lower	Lower	Lower	Higher
	2000	76.8%	77.6%	96.8%	281.4%	283.7%
	2008	80.4%	90.1%	100.8%	243.9%	309.8%
	% Chg	−4.7%	−16.1%	−4.1%	−13.3%	9.2%
	Troy	Higher	Lower	[ND]	Lower	[ND]
	2000	68.2%	118.0%	135.1%	206.9%	284.9%
	2008	124.4%	72.2%	[ND]	148.8%	[ND]
	% Chg	−82.1%	38.8%	[ND]	28.1%	[ND]
Earnings	United States	Lower	Lower	Lower	Higher	Higher
	2000	70.4%	85.3%	85.2%	90.5%	76.5%
	2008	77.6%	87.3%	87.0%	84.8%	70.3%
	% Chg	−10.2%	−2.3%	−2.1%	6.3%	8.2%
	MSA	Lower	Higher	Lower	Higher	Higher
	2000	74.0%	91.2%	87.9%	85.6%	85.9%
	2008	78.1%	84.4%	88.6%	76.2%	74.9%
	% Chg	−5.5%	7.5%	−0.8%	11.0%	12.8%
	Troy	Lower	Lower	Lower	Higher	Higher
	2000	84.1%	87.2%	75.3%	92.8%	63.7%
	2008	89.0%	90.4%	76.3%	85.5%	49.4%
	% Chg	−5.8%	−3.7%	−1.3%	7.8%	22.5%

Table 4.6

(continued)

Changes in Inequality, 2000–2008		White women as % of white men	Blk/AA women as % of Blk/AA men	Lat/Hisp women as % of Lat/Hisp men	Blk/AA women as % of white women	Lat/ Hisp as % of white women
Poverty	United States	Lower	Higher	Higher	Lower	Lower
	2000	128.2%	107.0%	114.7%	268.1%	265.9%
	2008	127.2%	118.8%	124.2%	258.3%	229.1%
	% Chg	–0.8%	11.0%	8.3%	–3.7%	–13.8%
	MSA	Lower	Higher	Higher	Lower	Lower
	2000	130.2%	112.3%	117.3%	390.2%	363.4%
	2008	121.7%	121.1%	133.3%	382.1%	361.9%
	% Chg	–6.5%	7.9%	13.6%	–2.1%	–0.4%
	Troy	Higher	Lower	Higher	Lower	Higher
	2000	110.0%	104.1%	137.9%	214.9%	318.8%
	2008	120.3%	75.7%	95.4%	168.0%	322.7%
	% Chg	9.3%	–27.3%	–30.9%	–21.8%	1.2%

Note: The 2008 figures are estimates based on the American Community Surveys, 2006–2008.

Change in Unemployment Inequality

In the United States, while women's unemployment rates have traditionally been higher than men's, the difference between the two virtually disappeared in the 1980s (DeBoer and Seeborg 1989). In the Capital Region, unemployment levels dropped for every group except Latinas or Hispanic women, who faced a 4.1 percent increase in unemployment between 2000 and 2008. White and Black or African American women closed the unemployment gap with men in their own racial category, reaching the dubious honor of equality in unemployment, while Latinas' or Hispanic women's unemployment levels far surpassed those of Latino or Hispanic men. Though the gap in unemployment between Black or African American and white women seems to be closing, the gap between Latinas or Hispanic women and white women actually grew by 9 percent between 2000 and 2008. In Troy, white women went from being unemployed 30 percent less than white men to nearly 25 percent more. However, Black or African American women's unemployment levels dropped relative to those of Black or African American men. In 2000, Black or African American women faced 18 percent higher unemployment than men in their racial group; by 2008, they faced unemployment 28 percent less often. As white women's unemployment grew, Black or African American women's dropped, closing the employment gap among women by race.

Change in Earning Inequality

At the regional level, earnings inequality is shrinking between white women and white men and between Latinas or Hispanic women and Latinos or Hispanic men, but is rising between Black or African American women and Black or African American men and between white women and women of color. In Troy, earnings inequality between women and men is shrinking, regardless of their racial group. However, earnings inequality among women by race is growing rapidly. For example, in Troy in 2000, Black or African American women earned only 7 percent less, and Latinas or Hispanic women 36 percent less, than white women. In 2008, by contrast, Black or African American women earned 14 percent less, and Latinas or Hispanic women 51 percent less, than white women.

Change in Relative Rates of Poverty

At the regional level, poverty rates are shrinking among white women and men. Among women by race, rates are roughly the same: Black or African American women live in poverty nearly four times as often as white women; Latinas or Hispanic women live in poverty roughly 3.6 times more often. Women of color are increasingly impoverished in comparison to men of their own racial group: an 8% increase for Blacks/African Americans and a 14% increase for Hispanics/Latinos. At the city level, while white women's poverty increased somewhat in relation to white men, rates for women of color shrank significantly in comparison to men in their own racial group, Black or African American women's poverty rates shrank relative to white women's, and Latinas' or Hispanic women's poverty rates rose slightly.

Being descriptive rather than predictive, these data do not definitively link rising inequality in the region to the arrival of the information economy. However, no flattening or leveling effect predicted by popular writers and municipal policymakers is shown. Moreover, the data suggest that high-tech development schemes like Tech Valley may increase economic vulnerability for women in the YWCA community, who, like many workers in Troy, suffer from legacies of inequality and discrimination in the area. Structural changes in the economy—technological change, globalization of trade, increased labor migration—seem to have combined with the neoliberalization of governance—withdrawal of the state from the

Already Here, Increasingly Vulnerable

Tech Valley initiatives may create an economy that increases the vulnerability of poor and working-class women—particularly those employed in the service industries. But what about the women in the YWCA community who were already employed in high-tech jobs? Did those women who followed the policy imperative to train for high-tech employment pull themselves out of poverty? Women struggling to meet their basic needs already perform high-tech labor of many kinds. But the experience of women in the YWCA community suggests that, even when they get high-tech jobs, these women continue to disproportionately face the negative impacts of high-tech employment.

In my years at the YWCA, I met three residents who were computer programmers by trade. All held advanced degrees, and one was a COBOL programmer who had excelled in computer science in the age of punch cards. In addition, one member of WYMSM was a systems administrator. If we relax the definition of IT-based work enough to correct for its innate class and gender bias by including data entry, insurance claims and processing, and telemarketing, more than a dozen more YWCA residents were engaged in high-tech labor. More than half of the residents I interviewed held jobs in data entry, call center customer service, telemarketing, telephone operating, and claims processing, and many of them had held several of these jobs over the years. Others identified significant computer and technological skills that were required for their work in the social service, secondary education, administrative, consumer service, and health care occupations.

Most of the women I interviewed who worked in high-tech jobs had contingent, part-time, or temporary work arrangements. None of their jobs offered health or unemployment benefits, most paid $7 or less an hour, and one rarely paid more than $50 a week for full-time work. The low wages, combined with the insecurity of the employment relationship in the high-tech sector and the physical and mental strains of the jobs, likely influence the extreme volatility and fast employee turnover in low-wage high-tech employment. For example, Cathy Reynolds was offered a promotion to supervisor after four months at an insurance claims company because of her seniority; only three other people had been at the company longer.

Call center and data entry work are sites of intense technologically-mediated surveillance, as the experience of women in the YWCA community amply illustrated. Work processes in call center and data entry provision of care, an increasing focus on individual self-sufficiency—create new configurations of inequality characterized by volatility, inse? rity, and precariousness.

Such complex forms of inequality cannot be explained—or remedied—in simplistic terms that pit women's economic interests against men's. A Guy Standing argues, "The era of flexibility is . . . an era of more general ized insecurity and precariousness, in which many more men as well as women have been pushed into precarious forms of labor" (Standing 1999, 583). Rather than illustrating women "catching up" to men in the region's high-tech workforce, the data suggest that work in the region is being feminized, both in the sense that a larger share of employment is going to women and in the sense that employment itself has shifted to have characteristics associated with women's labor force participation: lower pay, contingent and temporary work arrangements, and little or no job training (ibid.).[20]

Generalized imputations of "women's economic interests" and "men's economic interests," which underlie prescriptions to fill the high-tech pipeline with marginalized people or empower women to break into the boys' club of science and technology, overlook the feminization of work and obscure and marginalize important differences among women by race. In the last decade in the Capital Region, increasing equality among white women and men in education, earnings, and poverty rates has been offset by increasing inequality in these areas between women by race and among people of color by gender. In Troy, we see women of color beginning to catch up to white women in educational and employment achievement, but they face increasingly unequal wages and have enjoyed only minor success in reducing the astounding disparities in their poverty rates relative to those of white women. This is consistent with Leslie McCall's claim that the growing inequality among women by race is a defining justice issue in the information economy.[21] Women of color in Tech Valley are working harder, in more insecure conditions, for less.

I admire Troy's attempt to rebuild a flagging economy by becoming a part of regional economic development plans. But taken together, the indicators of existing racial, gender, and class inequality in Troy in 2000 and the new configurations of inequality established between 2000 and 2008 paint a fairly bleak picture of the city's potential to realize the benefits of investment in high-tech industry without exacerbating existing social stratification. In the absence of aggressive equity policies, regional high-tech economic development will not create broadly shared prosperity in Troy. Our investment in high-tech is thus an investment in injustice.

occupations are highly visible. Efficiency is measured and displayed constantly: employers track the number of bids entered or calls answered at the bottom of employees' computer screens, and check the numbers several times a day to monitor their progress. Target numbers are highly inflated. Francine O'Ryan was bypassing the work of other employees after six weeks on the job, sometimes entering as many as seventy bids a day, though the bids were often complicated—a hospital seeking bids for highly technical products such as drug testing kits, for example. Though Francine was prized for her quickness, employees' target was to enter 136 bids a day, almost twice the number of her best day.

IT systems in low-wage, high-tech work both centralize control and ask employees to constantly monitor and discipline themselves. Cathy Reynolds described the surveillance systems at an insurance company call center:

Cathy: They have the Internet there, but if you touch it they are right at your desk.
Virginia: How do they know what you are doing?
Cathy: Some monitoring system they have on your computer that goes back to the IT department. I mean they were like bloodhounds. I would be checking for some money in my bank account so I could go get some lunch . . . and they would be there, "Hi Cathy . . . How's it going?" And I'd ask, "Could I please just transfer money so I can go to eat?" They'd be like, "No, not on company time."

Information systems in the low-wage workplace distribute management responsibilities to workers—the running counter at the bottom of the screen—while simultaneously centralizing strategic control—the "bloodhounds" in the IT department.[22] Rather than fostering the kind of flexibility and autonomy for workers extolled by the high-tech management gurus, IT actually worked against the workplace self-determination of women in the YWCA community.

High-tech, low-wage employment is also fraught with adverse health effects. Women in the YWCA community complained that the work is both repetitive and physically demanding. Francine O'Ryan, for example, explained that she had developed cysts in her wrists, and had shooting pains in her forearms whenever she did too much data entry. Women also complained about the visual effects of staring at a computer screen all day. Often, women's health concerns are not taken seriously or are misattributed to other, psychological factors. Rosemarie O'Brian, for example, developed visual problems that made it impossible for her to use a CRT monitor. But when she told co-workers that she was unable to use a computerized sales register because of a medical condition, they insisted she was just afraid of computers and offered to teach her how to use them.

She described her interaction with the computer at her job in a department store:

Somehow the manager always understood that I had some [problem] with the computers and he never really made an issue of it. But one of the buyers took me [aside and tried to teach me to use it]. . . . He just thought I was afraid of it. I stormed off and said, "You just have to understand I can't do it." To this day he probably just thinks that I never learned how to use a computer, and that was my problem.

People's dismissal of Rosemarie's reluctance to use computers as "just in her head" extended to her interactions with many health professionals. When she went to a sports doctor for equilibrium problems related to her visual dysfunction, for example, he suggested she go to a psychiatrist. She explained, "He said to me, 'You just work in a department store. You don't do any physical labor.' I almost hit him over the head for that." This dismissal of women's health concerns in technological labor, or the insistence that their complaints are psychological rather than physical, has been well documented since the early days of the industrial revolution (Banta 1995; Killen 2006). Minimizing women's embodied response to workplace demands and discipline is a constant companion to efforts to rationalize women's work and bodies in the interests of social and workplace efficiency.

Though women in the YWCA community often found their high-tech work fulfilling, they faced many of the hazards of low-wage information economy employment, including job contingency and outsourcing, insufficient wages, little employer investment in training and reskilling, and health risks. To try to manage their increased "flexibility," many women I talked to mixed high-tech employment with more traditional service sector jobs, trying to balance the slightly higher wages of information industries with more stable and secure employment elsewhere. Several women in the YWCA community who had worked in data entry occupations left the industry to try to find more reliable work in personal services, child care, health care, and food preparation. The problem was not so much that they lacked access to jobs in the information economy but that the low-wages, employment volatility, adverse health effects, and disciplinary work arrangements of those jobs made the work unsustainable over the long term.

Holding Up the Pipeline

Women who hold these obviously high-tech—though low-wage, contingent, and temporary—jobs are not the only women participating in the information economy. Women holding jobs in the bottom half of the dual economy—social and health service assistants, teachers assistants, practical and vocational nurses, home health aides, orderlies and attendants, child-care workers, food preparation workers, cashiers and waitresses—are participating in the information economy as well. In fact, they provide the scaffolding on which the information economy rests by filling the gap in service and caregiving occupations that arise when the state neglects its obligation to protect individuals and families from the shocks and strains of the volatile economy.

Historically, women have disproportionately borne the costs of social reproduction, including childrearing, elder care, and community care-taking responsibilities. The costs of suspending a career path in informa-tion occupations to fulfill social reproduction responsibilities are particularly high, as employers offer few resources to make reskilling possible.[23] This is particularly vexing for women in IT employment whose careers have been interrupted by child-care responsibilities or the need to escape an unhealthy or unsafe personal relationship by leaving a geographic area to start over. For example, Francine O'Ryan described why her sister, who has a com-puter science degree, was unable to retain work in the high-tech sector. "My sister," she said, "has a degree in computer programming, and she is driving a bus. That's because her son is disabled. Every job that she gets, she takes too much time off. She is an aide on a bus, and here she is with a degree in computer programming. She's been out of it for so long that [her degree] is basically worthless."

Francine herself learned on the job. She advised other women who are trying to start a career ladder in the high-tech sector to take the risk of starting with a low-paying job in order to pick up skills. She remarked, "I hardly had any technology [skills] to begin with. I learned by stepping into different things, learning on the job. You get your foot in the door. Banks are the best places to start out at. They have all the updated tech. They just don't pay you that well." This is sound advice for women without primary responsibility for raising children or caring for the ill or elderly. But women with care responsibilities, particularly if they do not have a second income in the home, are often unable to take low-paying jobs in order to build a skill base. When mothers invest sweat equity, being paid too little in hopes of a big payoff down the line, they gener-ally invest in their children, and not in a high-tech startup. Juggling childcare responsibilities, living in insecure housing or a shelter system, lacking time and resources that would make it possible for them to mod-ernize their skills, many women are unable to reenter high-tech employ-ment. The mandate to self-train, central to the information economy,

Box 4.2
WYMSM Member Profile, Julia Soto Lebentritt

"It's great that we all think we have power, but we don't really unless we have economic freedom."

I moved back to Troy on New Year's Eve, 2000. Troy is my hometown, I grew up on the East Side. My dad immigrated to South Troy from Austria in the early 1900s. My husband and I came here from Burlington, Vermont, but before that I had been in Manhattan for fifteen years—quite a circular return for me. I remember seeing, from the back of one of my first apartments, the YWCA's Take Back the Night march. I thought it was marvelous, and the YWCA and WYMSM became important interfaces, connections to the community. Coming from New York City, I missed the opportunity to work with women from different cultures, different backgrounds.

I am interested in women, caretaking, and how we evolve out of the terrors of today. I am a lullabologist, which means that I study the role of lullabies in the transitional passages of a person's life. This work had been ongoing from the 1980s, when I started recording lullabies on site in New York City. Lullabies hit home for me because I am a poet and I know that musical language is extremely important in our lives, in our relationships. My book, *Following the Thread of Our Mothers' Joyful Caregiving: Lullaby-Based Activities for Caregivers of People with Dementia or Alzheimer's*, is scheduled for publication in 2009. What I have done is bring parenting techniques and traditions into caregiving for elders. As I encounter grief at the hospice where I work now, again and again I am finding that the lullaby is an important piece of relationship skills, bonding and mirroring. I feel a deep-rooted, heart-centered movement in my life to follow the lullaby, even coming back to Troy, meeting the women in the YWCA community.

I was introduced to WYMSM through the Women's Economic Empowerment Series, which I was attracted to because of the word *economic*. The empowerment movement has been ongoing, and has done wonders. It's great that we all think we have power, but we don't really unless we have economic freedom. That's still a big issue for women. I was an emerging woman writer in the 1970s; I was always on the fringe. I got free. The problem all along was, if you are outside the system with all your freedoms, and you're not married to a wealthy industrialist, how do you have the economic power you need and still pursue your creativity and your gifts, your talents?

The thing I liked most about WYMSM was the structure of the meetings. That collective approach appealed to me. We created our process, together. We struggled with it sometimes, but it was dynamic and people-centered. WYMSM achieved bonding between women of diverse backgrounds and

Box 4.2
(continued)

cultures in a common effort. We struggled together to create meetings where we all could vent, know and understand each other better, and find ways to support our individual paths. WYMSM integrated me back into downtown Troy. It gave me an interface with the community that carried through to later projects like "Sewing for Survival."

Five years from now, I want to see my book out. I want to see my vision through and have it move to other levels. I intend to be talking to a lot of different people. Five years from now, I hope to be one of a growing power-ful band of women making positive social movements throughout the world. I think it is important to take risks. This project was a risk. It is important to do something that no one else has done, especially if the motive is improving people's lives. Don't hold back too long. You have to move these things when you get them, and don't wait for the grant to come along. Take the step.

Based on a conversation in her home, May 25, 2009.

disproportionately disadvantages women, who continue to bear most of the social costs of care.

It is in this context that many women in the YWCA community turn to the service and caregiving sectors for employment rather than fitting themselves into the more lucrative but less stable information occupations. Women constrained by structural inequity—poor women, immigrants, and women of color—hold many of the jobs in the bottom half of the dual economy. These women are filling the social reproduction gap left by three historical forces: (1) the withdrawal of the federal government from its responsibility for caring for vulnerable citizens, (2) the admirable gains made by white and middle-class women in attaining education and profes-sional employment, and (3) the failure of men to step into caretaking roles traditionally borne by women. More privileged women are graduating from college and entering the professional and technical occupations at rates rapidly approaching those of men of their own race and class. But the success of these women—and the men who count on their social reproduction labor—relies on battalions of women working in service and caregiving positions, in industries that produce incredible vulnerability and risk for low-wage workers and their families.

High-tech equity will not be achieved by "fitting the poor into the new economy," the policy approach preferred by most national and regional initiatives.[24] They are already there in large numbers. The exploitation of poor and working-class people—especially women—makes the information economy possible. The distributive paradigm is at work in high-tech economic development schemes, like the Tech Valley initiative, that assume that people struggling to meet their basic needs simply lack technological training and opportunities for high-tech employment. Distributive solutions suggest increasing access to technology and technology training to bootstrap workers into the new economy. Empirically, this strategy is contradicted by the experiences of women in the YWCA community, many of whom have high-tech jobs of various kinds but are still unable to meet their basic needs. The attempt to fit the poor into the new economy also assumes that employment in the new economy will provide a better standard of living for workers. But available jobs in the information economy tend to increase poor and working-class women's economic vulnerability: the jobs held by women in the YWCA community were often unsustainable, exploitative, and failed to pay a living wage. The distributive paradigm in IT policy, therefore, underestimates the ways in which the information economy heightens risk for the economy's most vulnerable workers.

High-Tech Development in an Unflat World

The vulnerability of American workers, particularly those already marginalized by race, class, and gender, became increasingly clear as the global financial crisis touched more people's lives and brought the risks of the new economy to the doorsteps of middle-class homes. There is nothing intrinsic to the information economy that delivers the level playing field, flattened hierarchies, and increased opportunity promised by writers like Kevin Kelly or Thomas Friedman. The information economy does not sweep inequality away before it in a cleansing deluge. Rather, it injects more unpredictable, explosive change into an economic field already marked by durable disparity. The information economy is not Noah's flood, it is Hurricane Katrina.

On the Gulf Coast, while the hurricane itself was not entirely predictable, its effects certainly were, if attention was paid to the existing topography of inequality. It could be predicted—and was—that a natural disaster of that scope would have the most dire impacts on people living in valleys of inequality, like the Lower Ninth Ward in New Orleans, where geographic, historical, and political forces created disparities in infrastructure, resources, and political power. The hurricane itself did not create the disaster that followed. What created the tragedy was poor infrastructure, lack of effective emergency transportation, local policy more concerned about the economic viability of Bourbon Street than about justice for its citizens, and the federal government's unwillingness to protect the country's most vulnerable people. The combination of existing gender, race, and class inequalities and volatile weather resulted in dreadful consequences— death, displacement, and impoverishment—which were unjustly distributed along lines inscribed by enduring social stratification. Further, as Loretta J. Ross of *SisterSong* explained, "The hurricane and the subsequent flooding exposed the special vulnerability of women, children, the elderly and the disabled by revealing the harsh intersection of race, class, gender, ability and life expectancy. Many people could not escape not only because of poverty, but because they were not physically able to punch through rooftops, perch on top of buildings, or climb trees to survive" (Ross 2005, 1). Many more were unable to escape the devastation of the hurricane and flooding because they were unwilling to leave behind those who could not escape on their own. Such is the cost of social reproduction in a volatile world.

But just as the flattening of opportunity is not a natural or inevitable consequence of the information age, neither is increasing stratification and inequity. Rather than sit by and wait for the information age to magically deliver a level playing field, we must create thoughtful, rigorous policy that protects workers, individuals, and families who face more than their fair share of the burdens of the dual economy without garnering many of its benefits. We need to provide better policies, stronger levees, and feasible escape plans for workers on the front lines of the new economy.

5 Technologies of Citizenship

Virginia: Since you've been working with data entry lately, has it made you think any more about how DSS [the Department of Social Services] uses information systems?

Nicole: Just how *easy* it is to get the information. Who knows who has access. And who knows who I know back there. They have your Social Security number, your birthday, your mother's name, your father's name, your birth certificate, your Social Security card, your picture. Everything. They can ruin your life. Who knows? But at the same time, you have to keep the faith in them, that that won't happen. . . . When you go to the window and you tell them you want to see your worker they ask for your case number or your Social Security number. You hear a million social security numbers a day at DSS. When they punch it in, all my information comes up. I can look through the glass and see what my previous addresses were. All they have to do is get your Social Security number and they're in there.

Virginia: It's interesting how computers become the face of the system. . . .

Nicole: Yeah! That's just what it is, too—that screen behind the glass—all you have to do is tell them your numbers. They're going to stick a chip in us soon. [laughter] They'll just make you stick your hand through the glass, and they'll scan you.

Virginia: You and I should write a science fiction book!

Nicole: For real, though. It's happening. And it's not so much fiction. My mother says about immunizations—they used to just line us up and poke us, poke us, poke us. Now they're going to poke us with some chips. They're going to know *everything*.

—Nicole Thomas, interview, February 2, 2004

Seeking social and economic justice in an information age is not as easy as "a computer in every pot." Though access-based strategies like community technology centers are an important part of the high-tech equity agenda, the distributive paradigm has radically constrained the kinds of social and economic justice problems we are able to recognize and address. Because our understanding of the problem is so myopic, our universe of possible solutions and strategies for change is incomplete and restricted. One area that has been badly underexplored is the relationship among technology, the state, and citizenship. One of the most surprising insights from my years working on technological justice issues with women in the YWCA community was the ubiquity of IT in their everyday experience. But rather than offering them increased flexibility, mobility, and democratic input into the political systems that shape their lives—one of the great promises of IT offered to middle-class consumers—IT often increased their personal vulnerability, constrained their opportunities, and made them deeply suspicious of the political process.

Women in the YWCA community spoke at great length about the information-gathering and surveillance technologies—management information systems (MIS), electronic benefits transfer systems (EBT), closed-circuit television, and biometrics—in wide use in the social service system. Surveillance of poor and working-class women in the United States is nothing new, though the techniques may have changed. Conceptually, computerized information systems in the social service system are not very different from invasive home visits by caseworkers, extensive case records, or evaluations in workhouses by "overseers of the poor."[1] Politically, the purposes of surveying the poor have largely stayed constant for three centuries: containment of alleged social contagion, evaluation of moral suitability for inclusion in public life and its benefits, and suppression of working people's resistance and collective power. Practically, being lined up and poked for immunizations is not unlike having an embedded microchip scanned at the welfare office. Using marginalized people as test populations for technologies of state surveillance and control is not unique to the information age. From techniques of reproductive sterilization to methods of industrial psychology, the canaries in the coal mine of technological change have routinely been the poor and oppressed.

What is new, then, about the technopolitical experiences of women in the YWCA community? The women I interviewed perceived significant, and troubling, changes in the way that the social service system has operated since the integration of IT in the mid-1980s. Their critiques fell into three categories. First, IT has facilitated an intensification of surveillance and discipline. Welfare innovations like MIS and EBT cards make possible more precise tracking and monitoring of client behavior. These technologies act to significantly limit clients' autonomy, opportunity, mobility, and self-determination: their ability to meet their own needs in their own way. Second, as public assistance is computerized, the system becomes increasingly opaque, unpredictable, and arbitrary. The rapid sharing of database information between agencies lends credence to clients' fears that they are trapped in a system where every detail of their lives is known and freely shared among powerful players: caseworkers, employers, politicians, and police. Rules for information gathering, sharing, and retrieval are obscure, and mechanisms ensuring accountability are rare. Finally, IT systems and the specialized expertise that sustain them fragment the knowledge of social service clients and workers alike, misrepresenting the lives of people they seek to efficiently describe. The rigid architectures of new technologies of welfare administration do not allow for contextual information—the lived experience, struggles, purposes, and motives of women doing the best they can to survive and raise their children with dignity.

It is critical to ask what kind of social world—and what sort of citizen—IT produces when it is deployed by the state in the social service system.[2] Welfare administration technologies are political in two senses. First, they provide a form of hidden legislation, an emerging technological constitution that shapes manageable subjects for neoliberal governance and can play a significant role in reproducing power asymmetries. Second, interaction with these technologies provides moments of political learning for poor and working-class women, teaching lessons about the state, the operation of government, and the efficacy of making political claims.

An Emerging Technological Constitution

In his generative 1977 work, *Autonomous Technology*, Langdon Winner argues that technological artifacts, being the result of human design, necessarily embody specific forms of power and authority that encourage certain attitudes and values and discourage others. That is, artifacts have politics (Winner 1977, 19). Technological artifacts, Winner argues, can encompass political purposes far beyond their immediate use, and are often introduced into situations to resolve long-standing political questions or entrenched power struggles.[3]

For example, after the student unrest of the 1960s, the architecture of university and college campuses in the United States changed dramatically, shifting toward decentralized designs. Rensselaer Polytechnic Institute, where I received my Ph.D., was built in 1824 and features a broad grassy quadrangle in the center of campus that serves as a communal center of student life. By contrast, the University of California–Santa Cruz, where I received my undergraduate degree, and the uptown campus of the University at Albany, SUNY, where I now teach, held their first classes amid campus construction in the mid-1960s. UCSC has an entirely decentralized design that spreads independent colleges out among hundreds of acres of hills and woodlands. What common space exists is specific to each college, holds barely a few hundred students, and is separated from other colleges by long, narrow paths. The uptown campus of the University at Albany is physically separated by a nest of highways from the urban center of the city of Albany and features a sunken area arranged around a large fountain, accessible only by a few sets of narrow stairs, as the social center of the campus. Newer designs effectively contain students' potential political protest, making it impossible for them to gather in large numbers, align with others, or move quickly as a group.

Winner argues that technology must be understood as a form of legislation, part of an institutional structure that shapes policy and political life. Campus architecture, for example, shapes the social and political life of students in ways that are generally unacknowledged, and therefore its impacts are not open to democratic discussion and debate. Technological systems thus act as hidden forms of political and social order. While you can imagine that an explicit university policy banning gatherings of twenty students or more would be vigorously resisted, the spatial technologies of the new campuses pass largely unnoticed.

Lawrence Lessig makes a similar point in his 1999 book, *Code: And Other Laws of Cyberspace*, arguing that the onset of the information age has centralized the political power embedded in technological systems, despite cyberlibertarian rhetoric that claims that the Internet is a space free of social controls.[4] Like Winner, Lessig calls for expanded understanding of how regulation and other forms of political life are constituted through the architectures of our technological artifacts. He urges technology users and designers to condemn the libertarian premise that cyberspace can regulate itself and to consciously create a world where freedom can flourish through political-technical interventions such as open-code systems.

At issue for both Winner and Lessig is the lack of democratic deliberation in the emerging technological constitution. Because we so often believe that technology is self-directing, autonomous, and driven by inevitable progress, they argue, we give up our power as citizens to shape our destiny. Technological legislation is often written, by default, by scientists, engineers, and architects, who rarely include democratic principles in their research and design. As Lessig writes,

We build liberty . . . as our founders did, by setting society upon a certain constitution. But by "constitution" I don't mean a legal text. . . . I mean an architecture . . . that structures and constrains social and legal power, to the end of protecting fundamental values—principles and ideals that reach beyond the compromises of ordinary politics. Constitutions in this sense are built, they are not found. (Lessig 1999, 6)

We can create technologies that protect socially just values or we can build technologies that permit those values to disappear. We must actively choose the kind of technosocial worlds we want to inhabit.

My concerns with Winner's and Lessig's otherwise stellar work lies in their lack of faith in ordinary people's ability to understand and intervene in the emerging technological constitution. Both authors obliquely or directly accuse the public—excluding a few radical organic farmers and open-source software coders—of apathy, submissiveness, even complicity in existing technosocial arrangements. Winner argues that Americans' relationship to technologies of control is characterized by somnambulism; Lessig argues that "too many believe that liberty will take care of itself" (ibid., 58). But women in the YWCA community readily recognized how technology shapes their citizenship and opportunities for political mobilization, describing the ways that IT systems are used within existing political structures to track, monitor, and constrain their behavior. They articulated sophisticated critiques of how IT is being used in increasingly disciplinary ways, and made reasonable predictions as to the coming impact of the information age on their lives and livelihoods.

Political Learning in "the System"

Information technology (IT) helps forge a technological constitution through its deployment within state institutions. Positioned in existing systems of political and social inequity, IT shapes behavior, identity and political participation, provides opportunities for political learning, and teaches citizens lessons about government, democracy, and power. Interaction with IT in the social service office is a crucial form of technological engagement that teaches women in the YWCA community lessons about their role as client-citizens in a new informational order. Client status and interactions with the social service bureaucracy comprise an important part of poor and working-class women's political experiences, and those experiences are highly mediated by IT.

New work on policy feedback in political science offers important insights into how government institutions teach lessons about governance, insights that can be extended to help understand how technologies of citizenship operate politically.[5] Joe Soss, for example, notes that social assistance programs, such as Aid to Families with Dependent Children (AFDC) and Social Security Disability Insurance (SSDI), provide many beneficiaries with their most direct exposure to formal political institutions, and therefore shape political participation and learning.

"These programs," Soss argues, "provide the handiest and most reliable points of reference [for thinking about governance]. . . . Program designs not only communicate information about client status and agency decision making but also teach lessons about citizenship status and government" (Soss 1999, 376).[6] In fifty in-depth interviews with AFDC and SSDI recipients, Soss found that, through their program experiences, clients of AFDC—a means-tested program that requires intense and often invasive relationships with caseworkers—came to see agency decision making as an autonomous and unresponsive process, unconstrained by formal rules. Clients also saw that their degraded status "on welfare" put them in a position where asserting their grievances, even in situations with profound effects on their families and themselves, seemed both unprofitable and unwise.

Soss reasons that political lessons learned in the welfare office may be partially responsible for the much lower participation rate in formal political processes, such as voting, reported by AFDC recipients. In support of this theory, Soss finds that SSDI recipients—who do not experience an ongoing need to prove their eligibility for assistance, engage in casework relationships, or experience the social stigma attached to receiving welfare—have generally positive experiences with the social service system. Soss notes that their very different experience of the social service system may account for their more active political participation, which is roughly equal to that of similarly situated citizens who do not receive SSDI.

It is important to note that the AFDC clients Soss interviewed saw the welfare office not as a part of government but as a *microcosm* of government, so that "lessons learned about speaking up at the agency spill over into other forms of political demand making" (Soss 1999, 367–368). Soss writes,

As clients participate in welfare programs, they learn lessons about how citizens and governments relate, and these lessons have political consequences beyond the domain of welfare agencies. Program designs structure clients' experiences in ways that shape their beliefs about the effectiveness of asserting themselves at the welfare agency. Because clients associate the agency with government as a whole, these program-specific beliefs, in turn, become the basis for broader orientations toward government and political action. (364)

These orientations toward government and political action did not, Soss finds, stem from clients' low estimation of their own political capacity but rather from their estimation of the efficacy of making political demands.[7]

Box 5.1
WYMSM Member Profile, Cuemi Gibson

"All we've got to do is love each other, and care for each other."

What I would like people to understand is that we are all one person, one world. We're all human beings and we all have responsibilities to each other. People are hungry, and I watch as we throw away food. People are homeless, and I watch while they build parking lots. People are drug-addicted, and I watch how we allow drugs to fester in one area and not in another. We need the human factor: loving people, caring about people.

I grew up in South Troy with fourteen brothers and sisters. I'm a Black lesbian, and I like to think I'm self-made. I've always earned what I had, but I haven't had much. In the past, I've known the power of my intellect, but I didn't recognize it as well as I do now. I was afraid to be educated in a more mainstream, institutional way. Now I think I can combine my family background, my cultural history, and my traditions with mainstream education. It would make me more balanced, stronger, and more of a force for myself and the rest of the world.

The most meaningful thing WYMSM did was address how poverty and welfare were systematically shackling people, restraining them, dehumanizing them. But it also helped us understand that we were not alone. We had people from the outside come in and work with us, and it built our self-esteem and gave us something productive to do. The group gave members enough strength to get their needs met in a better way.

I was also excited by the learning technology part of it. When I went into WYMSM, I knew nothing about computers, and I had a fear of that because everything is automated now and there is no human factor. Now, I understand that technology can give me a faster route to get things I need. Technology's here. No matter how much I feared it, it has helped me and given me advantages that I didn't have before.

WYMSM taught me—technically and socially—that the world is not so fucked up. When I went in the military, I didn't think there was racism. I grew up in a neighborhood where I honestly believe there was color-blindness. But in the military, my drill sergeant was racist, didn't like me for who I was, because of my color, the pronunciation of my name. Prejudice is a dislike; racism is power. Somebody can dislike me and we can still have common ground! But if they dislike me and they have power over me, I don't have any level where I can fight back. I'm going to the highest entity of government, and I'm feeling racism! It made me standoffish and fearful. When I left the military, I went right into a homeless shelter. That says something!

Box 5.1
(continued)

When I went to WYMSM, I let my guard down. I *hate* welfare. I went on welfare and immediately got off because I didn't want to deal with it. Now I'm back in the same position, because socially, on an intellectual level, I changed, but emotionally I didn't. I never nurtured those changes. Now I'm getting to root again. I'm dealing with a lot of stuff being a Black woman and a lesbian. I've always indirectly fought racism and homophobia, but I've never directly dealt with them. I've always been alone in that fight, and it caused problems for me with jobs, housing, even my family.

Now, I'm addressing these issues directly, with other people. All this is coming full circle, and it's because I was in WYMSM. The women in WYMSM were strong, and we were very diverse—of different lives, colors, opinions, and standards of living. But we all meshed together as a group and came to some common ground. Our situations were different, but we had camaraderie, an understanding of each other. It wasn't always fun! But as a whole, we got something done.

In five years, I want to be an advocate. Being homeless at this time . . . I recognize the people around me. I want to advocate for younger people to prevent mental illness and homelessness. We have lost our children. Our education system has failed. Our mental health system is in denial. There's kids out there, twenty-years-old, homeless! Nowhere to go! And mentally ill! I have a lot of experience and understanding of this. I'm not a doctor, I'm not a social worker, but I've had lessons in my life, and one of those lessons was the experience of homelessness.

I have always wanted a utopia. They call it fantasy, but I don't think love is a fantasy, and it doesn't bear any race, color, or creed. My message to the reader would be, "All we've got to do is love each other, and care for each other." I don't believe in war. I don't believe in kicking somebody when they're down, or hurting people for no reason. We do that when we don't want to look at ourselves, at our pain. Sometimes, when someone smiles at you, smile back. When somebody needs to be embraced, embrace them. Extend to someone, if they need it. I just think we all should love each other, and that's my message.

Based on a conversation that took place in the author's home, November 25, 2009.

Technologies of State

Drawing on the insights of policy feedback, I sought in interviews and conversations to answer the question, what lessons about politics, government, and citizenship are women in the YWCA community learning from social service information systems? The women I interviewed expressed concerns about the ways in which IT provides a means of more efficient discipline and control of welfare recipients, reflected on how IT intensifies the lack of clear and accountable decision-making processes in the social service system, and bemoaned the absence of contextual knowledge about their lives and struggles. Their concerns fell into three broad categories:

1. IT is being used to build an increasingly invasive and disciplinary system of citizen control. Rather than discouraging fraud or increasing efficiency, IT largely serves to track and monitor individuals' behavior. Technological innovations act to limit clients' opportunity, mobility, and self-determination.
2. Welfare administration technology seems increasingly opaque, unpredictable, and arbitrary to its clients. Rules for information gathering, sharing, and retrieval are obscure, and mechanisms ensuring accountability are rare.
3. IT systems and the specialized expertise that sustain them extract and fragment the knowledge of social service clients and workers alike, misrepresenting the lives of people they seek to describe. Rigid IT architectures shear away context and limit the possibility of attaining a holistic viewpoint or critical coherence that could help clients meet their needs in their own way.

Women in the YWCA often perceived IT as threatening, intimidating, and extractive because of how personal information is used against them in the bureaucratic information systems to which they are subjected. They drew a realistic sense of threat from these experiences, attributing an invidious agency to the technology.

"The computers find out who you are, too," Cuemi Gibson argued, "Because I'm sure when they put your name and your Social Security number in there that everything comes down. That's my experience with SSI [Social Security Insurance]. Here you go—all the way into the system now. Every part of your life, everything about you, is available. . . . The system—that person that you're talking to—knows everything about you. Knows more about you than you know about them, and that's not a fair game." This perception of IT as threatening and invasive cannot be attributed to simple fear of technology or resistance to change. Gibson's intuitions about the role of IT systems in the department of social services were more than adequately confirmed by the experiences of other women in the YWCA community.

Tracking and Monitoring Behavior

Women in the YWCA community expressed concerns that recent innovations in social service technology—the distribution of benefits on EBT cards, for example—facilitate more precise tracking and monitoring of client movements and behavior. EBT cards were purportedly introduced to minimize welfare fraud and reduce the social stigma of using food stamps. As Veronica Macey explained, "They don't give you the little books of food stamps anymore—it's a little credit card. [They changed that] because people were selling their food stamps to drug dealers, [but] people still do it all the time. They just take their drug dealer grocery shopping with them." Dorothy Allen concurred: "They did it to prevent people selling their food stamps and stuff. But they can still sell their food stamps! . . . I give my card to my mother all the time—so if she needs anything, she can just go get it—if [fraud] is what they're trying to prevent, they didn't find the solution."

In the estimation of women in the YWCA community, the EBT card is a remarkably ineffective method for curtailing fraud. Recipients have simply become more creative in finding ways to survive in the system. My interviewees suspected that eliminating fraud was just a cover story for the cards' actual purpose, tracking welfare recipients' movements and purchases. Dorothy explained, "I think [it's for] tracking, because there's not only food stamps on the card, you get cash, too. Like me, my babysitting money comes on my card—I give my card to my babysitter, have her take it all off. [Once,] she ended up going to the mall, and she's swiping it and swiping it. So [the next time I had an appointment at] DSS, they asked me, 'Why is your money being taken out like this?'. . . They use these cards as a tracking device. That's what it is."

Limiting Options and Mobility

Concerns about lack of privacy and agency abuse of information keep many people from collecting their entitlements. As Cuemi Gibson argued, "A lot of people won't go to DSS because of the privacy aspect—the information aspect—of it. You all know who I am and everything I do! And you only allow me $290 [a month for rent] and then won't let anyone live with me. Why can't I have a roommate? This money—you're saying it's mine, but you're monitoring what I'm doing with it. That's not fair!" Therefore, in a very real way, IT supports a system that limits the options and the freedom of women struggling to meet their basic needs by keeping them from attaining the support to which they are entitled by law. It also constrains clients' behavior in more subtle ways: because benefits are now distributed on cards, clients are unable to shop at many small local stores and farmers' markets, which lack card readers. Their benefits card thus limits their nutritional choices and acts as a tacit endorsement of more expensive and often less accessible chain stores.

Lack of Transparency

The opacity and complexity of IT systems cause further problems. "It just seems like," Cathy Reynolds explained, "from person to person the rules fluctuate. I don't see consistent anything. Because I look at me and Theresa. She is pregnant and I'm pregnant but they won't give her cash assistance. They won't help her out. . . . I don't understand any of it. I don't know if [rules] in the computer [determine if] people get denied, and that's how they do it, or what their basis for accepting and not accepting are. . . . [At DSS], they write their notes and then they disappear, and then they are like, 'Alright, give us a week.' Do they just look at it and are like, 'OK, data, data, data'? Or is it like fields on the computer screen you have to enter, and then the computer goes through all the rules that it has been pro-grammed with, and [then it] says 'Accepted' or 'Denied'? How is it based? Because it makes no sense to me. . . . I get my $139 of food stamps. I get my $62 every two weeks. Yet Theresa doesn't. From one person to another it just switches. . . . [Virginia: "It seems just like a random lottery?"] . . . Right. You get the special prize. You get money."

The feeling that rules governing public assistance are arbitrary is intensified when benefits are electronically distributed. It seems like magic, women in the YWCA community explained, benefits just turning up on or disappearing from your EBT card. Dorothy Allen laughed and said that it seems like her phone number receives her benefits, not her. Reynolds remarked, "The food stamps and all that. They put them on your card on certain days. I would die to find out how to be the first day or the second day. I have to wait until the eighth and I can't wait until the eighth. It is too far away. I'm starving—I want some food." Veronica Macey considered the effect these issues have on women's relationship to technology. "I know that a lot of people get mad at the computers at social services, because sometimes they won't put their food stamps on their card on time, or their cash benefits. So they are calling DSS all the time, asking 'Where is my money?' And they may have no food. I know a lot of people get really mad about that."

Undisclosed Information Sharing (Information Abuse)

Federal, state, and community agencies routinely share database information among themselves, which lends credence to clients' fears that IT creates an all-pervasive system of surveillance and control. When requesting any kind of assistance, clients in New York State must sign a "use and disclosure statement," which allows the Department of Social Services to share the information they provide on intake forms about themselves and their family members with U.S. Citizenship and Immigration Services and with "people and organizations directly connected with . . . the administration or enforcement of the provisions of the Temporary Assistance (TA), Food Stamp Benefits (FS), Medical Assistance (MA), Medicare Savings Program (MSP), Child Care Assistance (CC), Foster Care (FC) and Services (S) Programs."

Because each program has a separate staff of caseworkers, examiners, and administrators, and because neoliberalism and privatization have resulted in the outsourcing of many of the duties of public assistance programs to nongovernmental organizations, client information is accessed by dozens of widely dispersed and largely invisible people and agencies.[8] The 1996 Personal Responsibility and Work Opportunity Reconciliation Act both limits lifetime benefits to sixty months per individual and demands that social service offices across the country computerize. This dual mandate requires that data be available across state lines and be retained for a client's entire life. In addition, the Department of Social Services practices "number matching" every time a client applies or is recertified for assistance. DSS runs the individual's Social Security number—and, increasingly, her fingerprint—through a series of state and federal databases to confirm her identity; verify information about employment, earnings, assets, and child support payments; search for outstanding arrest warrants; and gather other information.

Few clients remember signing the use and disclosure statement, and when asked if their caseworkers discussed with them how their information is used or their rights to confidentiality, the great majority say no. Though it is clear that they have been advised that they have no reasonable right to privacy, life circumstances and institutional structures shape their ability make choices to disclose or withhold personal information. The public assistance intake packet in Rensselaer County is 109 pages long, and the use and disclosure statement appears on the twenty-third page of the packet. Most people trying to access public assistance are in dire emergency situations and under a great deal of stress. Clients can refuse to sign, or refuse to supply a Social Security number. But this would disqualify them from receiving any kind of assistance, or even from speaking to a case-worker. If clients do not sign away their right to control their own information, they will not receive any kind of benefits: food stamps, transportation vouchers, Medicaid, housing assistance, child care, emergency cash assistance, and other basic necessities of life. Under these conditions, notions of "free choice" and "informed consent" are stretched to their breaking points.[9]

Even when client confidentiality is specifically protected, as in the case of domestic violence victims[10] or youthful offenders, there is little knowledge of how, or if, the information is actually safeguarded. This lack of transparency leads to the abuse of information in local agencies and to the widespread, though incorrect, belief that there are no safeguards for client confidentiality at all. YWCA staff member Gina Lancioli, for example, expressed concerns about rampant, inappropriate information sharing: "A lot of women . . . face that at the public housing authority. They can't get in [to public housing] because of something [in their past]. . . . [T]he crazy thing is that [the housing authority] is not supposed to be able to look at records from when you are a teenager, but they find out." Under pressure to sign away their privacy in exchange for benefits, and rarely informed how their information is being shared, clients often feel that they trade away their rights, and every detail of their lives, in exchange for the basic necessities of survival.

Extraction and Fragmentation of Knowledge

Information plays a complex role in women's poverty and the state's response to their demands. Complex disciplinary technologies collect and manage enormous amounts of data on women and men struggling to meet their basic needs. Still, women in the YWCA community did not reject the idea of information poverty outright, arguing that there are important informational resources to which they still lack access. In my July 2003 interview with WYMSM member Cuemi Gibson, I explained some of the ways that the concept of information poverty was being used in public policy, and asked for her reflections. She linked her complicated relationship with information and information technologies to her continuing experiences with racial injustice:

Virginia: Is the problem that poor folks are information poor? What do you think?
Cuemi: Well, yes, because we don't know our roots. We never did. We never were allowed to. Basically, I can't know who I am or where I'm from because our race

was scattered. Honestly. When I was growing up, my foreparents didn't know how
to read or write, so we didn't have a head start in that, Virginia.

Virginia: But if poor people are information poor, why does it take two and a half
hours to fill out DSS intake forms?

Cuemi: Because of the kind of information they're looking for. I recently got my
file from [a local agency]. People of color don't know that you can take back your
information. They don't tell you a lot of things. If you don't have an education, it's
hard for you to know how to seek out [even your own] information. We were taught
to believe that white people are powerful. My aunt used to say, "Don't fight people,
Cuemi." Because they thought it was threatening. Because of their foreparents'
experiences with white people. So the information that we had—our foreparents
wouldn't speak it. They were afraid—afraid to confront white people. So when you
say information poor, that's from history . . . and add to that that most people of
color are only getting a seventh-grade education, so we have two generations of
people who are walking around functionally illiterate.

Virginia: Is it as much a question of feeling that you lack control over your
information?

Cuemi: Yeah. People are systematically abused. . . . Like CPS [Child Protective Ser-
vices] has so much information on families. That's overwhelming, because they're
not searching for anything positive.

Gibson pointed out that it's not lack of information that matters—it's
control over information. For African American people, she argued, it is
risky to either volunteer or withhold information. She explained that in
the case of communicating with social service agencies, it is dangerous to
withhold information: "With [a local drug rehabilitation agency] when you
say 'I don't know, I don't know,' it works against you because they think
you're trying to bullshit them. So they automatically take you for a piss
test. And if you fail, your welfare income is denied. *So if you refuse them
information you lose your income.*"

On the other hand, it is also dangerous to *volunteer* information, particularly with the police. She explained that letting police know that you
are aware of your rights as a citizen can escalate an already tense situation.
In our interview, I asked if she felt that her being in control of her own
information made her seem like a troublemaker. She responded, "Absolutely.
Because I'm a Black woman." Though she acknowledged that sometimes
withholding information could be an expression of power, she insisted that
"the less information I have, the more power they have."

Loss of Context

It is not only social service clients who sometimes feel trapped by the
dictates of IT systems. Information technologies sometimes work against

the best intentions of social service workers. In a 2003 interview, a YWCA community member related a story about a job she had doing data entry for a contractor who was developing a tracking system for young people who were under state supervision. The frustration that finally drove her to quit the job was that the architecture of the database didn't allow social service workers to include narrative information about the context of kids' behavior. Simply, the system tracked each student's "success" or "failure" in a number of different programs. So, for example, if students stopped going to an afterschool program because they faced a serious crisis—a death in the family or an apartment fire, for example—a case-worker was forced to check a box that reported that they failed to complete the program. Because there was no input box for narrative case notes, there was literally no place in the system to account for the (sometimes pages of) contextual information written in the social workers' reports.

YWCA community member Rosemarie O'Brian narrated a similar experience in the mental health system, where she felt that her experience and knowledge were not being heard by a doctor infatuated with a new diagnostic system. "When you go into the mental health world," she explained, "there is this standard test which is called the 500 questions test . . . this has been one of the fundamental pieces of diagnosing people for a long time. So [I went to take it], and this guy had just transferred it to a computer program. . . . He was [so] completely overcome with delight that he could make this computer program . . . [that] he wasn't listening to me. I was so humiliated. He wasn't trying to help me figure out what I was dealing with. He was totally focused on his computer program." The problem, as these two community members saw it, has to do with what is interpreted as signal and what is interpreted as noise in an informational system. The "signal" is only what the database makes room to record—checkmarks that represent success and failures, the diagnostic question that is currently on the screen—and the noise is whatever experience, knowledge, or concern does not fit into these limited parameters. Both women argued that it is not necessarily the information that is collected about clients or patients that matters but rather what is left out. The structure of technological systems can erase the contexts and knowledge of people those systems seek to describe.

Lack of Critical Coherence

The women in the YWCA community defined the problem less as a lack of information and more as the fragmenting and specialization of knowledge—facilitated by IT—that leads to the elimination of more holistic and critical views of the world. "People," Rosemarie continued, "are losing something ultimately that relates to reason, to a lack of cause and effect, which ultimately leads to justice. . . . [The problem] with digitally related information...is that there is information that is known [but not used]. . . . It's like famine. There are tons of food; it just doesn't get where it needs to go." What is missing, she notes, is not information itself but an overview, the whole picture. In this insight, Rosemarie echoes Paulo Freire.[11] For Freire, critical literacy takes as its goal more coherent understandings of the world. Therefore, specialist knowledge has low coherence and little criticality. Critical literacy, on the other hand, fosters linkages between "self-contained areas of expertise" and the "social and political realities" that frame people's understandings and their integration of their ideas into the world.

Technologies of Citizenship

Different targets of technologies of state administration, like different target populations for public policy, receive different messages about governance and citizenship. There are strong pressures for public officials and IT designers to provide beneficial policy and systems to enable greater flexibility, transparency, and mobility for powerful, positively constructed target populations such as middle-class consumers, traditional nuclear families, homeowners, and racial majorities. Similarly, there are strong pressures to provide negatively constructed target populations, such as Temporary Aid to Needy Families (TANF) recipients, single parents, the poor and working-class, and racial minorities, with policy and IT systems that fragment knowledge, demobilize collective thinking and action, monitor and discipline behavior, and obscure the operation of political and bureaucratic systems (Schneider and Ingram 1993). The differential construction of target populations for IT systems, as for public policy, teaches lessons about citizenship, value, and the operation of the state.

The social service system provides many women in the YWCA community with their most direct exposure to high-tech information systems. It therefore provides the most common form of technology training that these women receive. Through their program experiences, women in the YWCA community learned to see the deployment of information systems as an invasive, autonomous, and unresponsive process, unconstrained by formal rules and unconcerned with transparency of process. Their degraded status in the system puts them in a position where controlling their own information—whether choosing to volunteer it or to withhold it—is difficult and potentially dangerous. In addition, my interviewees saw IT not simply as a tool of government but as a *microcosm* of government—information technology, for them, is the face of the system. Therefore, to modify Soss's argument, lessons learned about interacting with IT at social service agencies are likely to spill over into other forms of technological engagement and other forms of political demand making.

Rather than being unaware or apathetic, women in the YWCA community explicitly recognized their role as a test population for technologies of social control. In addition, they expressed real concern that the techniques and technologies used to regulate their behavior in the social service system would eventually be used on the population at large. Dorothy Allen, for example, explained that rich people were too insulated and naive to understand that the technologies that were tested in the welfare office would eventually be used on them. Women in the YWCA community often saw themselves as canaries in the coal mine, and therefore felt a high level of responsibility for political action and education on behalf of the general public.

Different technologies—social service programs and information systems alike—produce different forms of citizenship. Soss concludes his article about participation in social assistance programs by explaining that participation in more democratic programs like Head Start mitigated or superseded the demobilizing effects of AFDC. He argues,

More participatory program design encourages more positive orientations toward political involvement. Head Start provides clients with evidence that participation can be effective and fulfilling. From the perspective of participatory theory, it is not surprising that these experiences have spill-over effects. "The taste for participation is whetted by participation." (Soss 1999, 374)

As Winner reminds us, "Different ideas of social and political life entail different technologies for their realization. One can create systems of production, energy, transportation, information handling, and so forth that are compatible with the growth of autonomous, self-determining individuals in a democratic polity. Or one can build, perhaps unwittingly, technical forms that are incompatible with this end and then wonder how things went strangely wrong" (Winner 1979, 460).

We need to build different technologies of citizenship. But in order to do so, we must jettison the thinking that binds us to the distributive paradigm and narrows our vision of social justice in the information age. Within the confines of access-only approaches, it is impossible to acknowledge or understand the experiences women in the YWCA community have with IT in community institutions, the social service system, and the low-wage workplace. It is impossible to recognize, and therefore transform, the real world of information technology.

6 Popular Technology

I think the poor . . . have a sense that without structural changes nothing is worth really getting excited about. They know much more clearly than intellectuals do that reforms don't reform. They don't change anything. They've been the guinea pigs for too many programs. Now if you could come to them with a radical idea . . . they'd identify with that, but not with short-range limited objectives that they know from experience don't get them anywhere. They won't invest much time or energy in it.

So to embolden people to act, the challenge has got to be a radical challenge.

—Myles Horton and Paulo Freire, *We Make the Road by Walking* (1990, 93–94)

The stories women in the YWCA community tell about their experiences in the information age can be disheartening. Based on the first half of this book, it would be easy to dismiss information technology (IT) as just one more thing that disadvantages people struggling to meet their basic needs in the United States, and indeed across the world. I hope instead that I have begun to explain the critical ambivalence women in the YWCA community have about the information age and to reframe this ambivalence as an opportunity and a resource rather than a barrier. The second half of the book explores what can be done to foster critical technological citizenship and engage *all* people in a broader discussion about the relationship among technology, citizenship, and social justice.

This chapter describes WYMSM's attempts to combine libratory education, research, and design techniques—such as popular education, participatory action research, and participatory design—with a commitment to high-tech equity at the YWCA. In it, I describe three projects: the construction of a community technology laboratory, the creation of the online Women's Resource Directory, and the design of a game called Beat the System: Surviving Welfare. These projects are examples of popular technology, an approach to critical technological citizenship education that takes advantage of the opportunities offered by critical ambivalence to empower participants to make transformative social change in their lives, their communities, and the world.

Coming to the YWCA

My connection with the YWCA of Troy-Cohoes was central to the creation of this new approach. Returning to Troy from a semester spent as a visiting graduate student at the Massachusetts Institute of Technology, I was interested in developing asset-based community development systems for sharing resources across nonprofit organizations in the community, an approach I had learned at MIT's MediaLab. I resisted early suggestions that I focus my research on only one organization, especially if that organization was the YWCA. The image I had in my mind of the YWCA was of an organization of Christian ladies in white gloves pouring tea and organizing charity drives. Convinced to go ahead and give it a try by Rensselaer faculty member Ron Eglash—I was interested in working with adult women and the YWCA was right across the street from my apartment—I originally understood my role at the YW in the limited, short-term sense of helping the organization develop technological capacity.

My mind was quickly changed by a series of meetings with YWCA executive director Pat Dinkelaker and several trips to the YWCA's weekly community meal. In meetings with Dinkelaker, I learned that the YWCA was not quite what I imagined. She and Sally Catlin Resource Center director Christine Nealon had recently returned from a Ms. Foundation Institute for Women's Economic EmPOWERment and were full of ideas for combining Rensselaer's interest in developing community technology resources with their mission-based goals of economic empowerment[1] of women and girls and the elimination of racism. The two women introduced me to the mission of the YWCA, which reads:

The YWCA of Troy and Cohoes is an affiliate of the Young Women's Christian Association of the United States of America, a women's membership movement nourished by its roots in the Christian faith and sustained by the richness of many beliefs and values. Strengthened by diversity, the Association draws together members who strive to create opportunities for women's growth, leadership and power in order to attain a common vision: Peace, Justice, Freedom and Dignity for all people. The Association will thrust its collective power toward the elimination of racism wherever it exists and by any means necessary.[2]

Expecting ladies in white gloves, I was surprised and intrigued by the ferocity of the YWCA mission, its focus on women's power, and the inclusion of what the organization calls the "one imperative," the call to end racism "wherever it exists and by any means necessary." The mission, Pat and Christine explained, provides a frame and focus for all the activities of the YWCA: the staff consciously used it to set agendas, mediate conflict, and develop collaborative programs with the ninety women who make their home in the building.

Dinkelaker steered me to resources in the literature on popular education, pointing me to Paulo Freire's *Pedagogy of the Oppressed* (1997), the programs of the Highlander Research and Education Center, and Petr Kropotkin's *Mutual Aid: A Factor of Evolution* (1902). At our second meeting, Pat gave me a printout of the following quotation from Freire's classic work:

Dehumanization is the result of an unjust order that engenders violence in the oppressors, which in turn dehumanize the oppressed. Because it is a distortion of being more fully human, sooner or later being less human leads the oppressed to struggle against those who made them so. In order for this struggle to have meaning, the oppressed must not, in seeking to regain their humanity (which is a way to create it) become in turn oppressors of the oppressors, but rather restorers of the humanity of both. This, then, is the great humanistic and historical task of the oppressed: liberate themselves and their oppressors as well. (26)

Dinkelaker let me know in no uncertain terms that any activities I undertook at the YWCA had to follow this spirit. Any activity I brought to the YWCA, she and Nealon insisted, had to be consistent with both the mission and the one imperative and participatory at every level and at every stage. They required from the beginning that all workshops and classes be co-facilitated by residents of the YWCA, that all activities provide leadership training, and that technology programs be designed so that community members felt empowered to voice criticism and redesign programs themselves. They also insisted that university faculty and students understand at every point that the process we were embarking upon was an *exchange*. We were not simply giving our skills and knowledge to women at the YWCA. Women in the YWCA community were contributing something valuable to us as well: their time, attention, ideas, and labor.

Dinkelaker and Nealon piqued my interest in the YWCA, but the community meal cemented my commitment to working with the organization. The community meal is a YWCA institution. Every Thursday, residents and staff of the YWCA make dinner for everyone in the building and their guests. Held in the gym, the community meal regularly attracts a hundred diners and acts as a major community bonding ritual. In the early days of my relationship with the YWCA, community meals not only taught me

Box 6.1
WYMSM Member Profile, Christine Nealon

"My goal is not charity, it is organizing."

I come from a privileged background. I haven't had every privilege in life, but very few times have I worried about food, clothing, or shelter. Nevertheless, I was raised to think about how my actions affect other people, and to know that everyone's life was not like mine. I've always had a perspective outside of myself. I have a vision that includes a greater level of awareness for people of how their everyday actions affect peace in the world.

WYMSM was an organically built group of people that came together for different reasons, and together we learned and discovered things about ourselves, each other, and how we fit into the world. We all came with different agendas, but they weren't so strong that they overshadowed the goals of the group. I don't think any of us knew that we were going to create Hunger Awareness Day when we first started. We embraced everyone, acknowledged what our strengths and weaknesses were, shared resources, and ended up with a public event where people who traditionally wouldn't have been speaking did just that. People who held strong views about public assistance and poverty learned a lot from us—our county legislator, probation officers. It grew from a small group, bigger and bigger, beyond even what we could have imagined. It was like a train that was gaining momentum down a hill. We couldn't have stopped it if we tried.

I wish there was a way for the spirit of WYMSM to be integrated into the everyday workings and goings-on of social service organizations. Sometimes people see this work as extra "fluff stuff." But I can see a stark contrast between what we did and other initiatives for eliminating racism or empowering disadvantaged individuals. Standing up and saying "I'm against racism" is very different from sitting in a group of people listening to others' childhood experiences with racism, and then talking about one's own childhood experiences with privilege. Conversations like this help you understand racism on a deeper level and to take ownership over how you will continue to live. If we don't acknowledge the root causes of disempowerment, and we only talk about poverty as personal flaws or lack of skills in individuals, that approach doesn't work. Dealing with root causes is a whole different ballgame. My goal is not charity, it is organizing.

WYMSM helped me evolve. I went into that experience person-centered, seeing the effects of poverty on individual women. The YWCA and WYMSM started to reinforce my suspicions about the impact of structural forces. It reinspired me to continue this kind of work with women, with all different kinds of people. It can be overwhelming, but in the end it gives me so much

Box 6.1
(continued)

energy. In five years, I hope to have more balance in my life so I can continue to grow personally and professionally. Of course, I want my family to be happy and healthy, but most of all I want to look back on this day and know that I have continued to learn about life.

I want readers to know that the members of WYMSM were very concrete people, not people who started out with an impossible dream. I want people to know that we used tools like respect, listening, transparency, and resource sharing. We were open to learning and making mistakes, and great things came out of it. It can be done; it can be replicated. WYMSM was incredible. The connections that we made have moved me forward, and all of our paths have continued to cross. WYMSM didn't go away, it has transformed; it is living on in other ways.

Based on a conversation that took place over lunch, May 6, 2009.

more about women in the YWCA community but also forced me to question the limitations of my own understandings of the nature of poverty in the United States.

I went to my first community meal on September 27, 2001, the day after my first meeting with Dinkelaker and Nealon. In my field notes, I wrote,

Went to community meal with Jes [Constantine] this evening—found myself being quite shy, unsure of how to act or how to open conversation. Residents really span demographic categories here—from very old to very young; African American, Latina, white; from very transient (the day before, there had been nineteen intakes) to very moved-in (some residents have been there eighteen to nineteen years). Interesting how much more adept at dealing with the residents Jes is than I am. She's been here long[er], but I am a little ashamed at myself, that I wasn't able to just jump in with both feet. Maybe it's best that way—I don't want to overwhelm myself or anybody else. Or maybe it's just my own class and race blinders in full effect.

In retrospect, I see my surprise at who was in the YWCA community and my hesitation to fully engage residents as indicators of a deep, socially embedded ignorance. Though I knew intellectually that economic hardship wears many faces and that there is no singular culture of poverty, at some level I was not expecting the diversity of people living at the YW.

I thought that, by attending the community meal, I was reaching across a divide to people who were not like me. But when I got across the street and sat down to dinner, I had to face the challenging proposition that the women in the YWCA community were not so different from me after all. The mismatch between what I expected and what I experienced made me feel shy, and a bit ashamed. The community meal challenged my ideas about race, class, gender, and privilege, my picture of myself as unique and deserving, and the deeply embedded, unconscious cultural assumption that poor and working-class people are not "like us," that they get where they are because of innate deficiency or bad choices, or that they are lazy, unintelligent, downtrodden, and apathetic. At the community meal were lively, varied, engaging women. Small wonder I was uncomfortable! I was confronting, in the empirical flesh, my sanctioned ignorance about poverty in the United States.[3]

Popular Technology Is . . .

Popular technology grew out of the spirit of the YWCA; it is rooted in a passionate commitment to ending injustice, meeting people "where they're at," and creating spaces for meaningful exchange born of mutual respect. Popular technology is an approach to critical technological citizenship education based on the insights of broadly participatory, democratic methods of knowledge generation. While it is informed by the lessons of social movements across the globe engaged in popular education, participatory action research, and participatory design, the technique grew out of a very particular time and place, within an organizational context dedicated to human rights, nurtured by the insights and input of women who lived, learned, worked, and played at the YWCA.

Popular technology creates a space in which all participants can become more critical technological citizens. It has as its goal the creation of collective knowledge, the practice of people's science, and the exercise of power in political movement. Popular technology is, therefore, a problem-posing rather than a problem-solving strategy for achieving equity in the information age. Rather than seeing adult learners as skill or knowledge deficient—as problems to be solved or empty vessels to be filled—popular technology assumes that most people have vast experience with IT in their everyday lives, experience that provides them with a substantial knowledge base and a critical desire to know more and understand more fully.

The goal of popular technology is to help everyday experts from a wide variety of social locations become more critical in their thinking by posing contradictions and problems in ways that lead them to the next stage in their analysis of the information age. To do this, popular technology draws on three related techniques: popular education, participatory action research, and participatory design.

Popular Education

Popular education is a reflexive, nondoctrinaire approach to collective, participatory learning. It inspires insight by building on what emerges from the everyday life experiences of participants in workshops or informal classes. My understanding of popular education stems from three primary sources: Jane Addams and the Settlement House movement, Myles Horton and the Highlander Folk School tradition, and the South and Central American work of Paulo Freire and the educators he inspired.[4] The underlying philosophy of these three movements is that knowledge grows from social experience and critical reflection on the conditions and contradictions of one's life.

Popular education trusts in the oppressed: in their ability to reason, analyze their experiences, and intervene in the dominant social order. Popular education is "popular," then, in two senses: in the sense that it is education carried out by and for people themselves outside of formal institutions, and in the sense of supporting the rights and powers of common people in their struggle to liberate themselves. Addams, Horton, and Freire are in a tradition that insists that democracy be the method and practice, and not merely a goal, of education. As Horton wrote in the early 1970s,

Education is too important to be left in the hands of institutions and experts. Decision-making must not only be decentralized; it must be pluralistic. . . . Efficiency will definitely not be the objective; nor will neatness or speed. But, given *bona fide* decision-making powers, people (as we have learned at Highlander) will rapidly learn to make responsible and socially useful decisions. Moreover, they will be quite willing to assume responsibility for carrying out the decisions based on their collective judgment. The problem is not the danger of irresponsibility or inefficiency, but the problem of convincing students, minorities, and other disenchanted people that their involvement will have meaning and their ideas will be respected. (Horton 1972, 31)

Popular education is learning about decision-making processes by actually *making* decisions.[5] It is the right of every person to have their full powers released by a system of mutual education that (1) has deep relevance to learners' everyday lives, (2) helps learners see their place in the world in relation to social structure and other people, and (3) is grounded in conscious training for action.

Participatory Action Research

Participatory action research (PAR) describes an orientation toward research that promotes concrete change by generating and assisting ordinary people's analysis and action. PAR "break[s] down the traditional boundaries between knowing and doing [and] put[s] people traditionally thought of as 'research subjects' in charge of investigating and transforming their own world" (Williams 1997, 1). PAR often combines grassroots research projects with popular education and direct action organizing (Reardon 2000) in order to provide more equitable control over the means of both material and intellectual production. In so doing, PAR creates reliable and rich empirical knowledge of social conditions generated by the people who most directly experience them, provides space for the growth of critical collective self-consciousness, and mobilizes people to achieve transformation of social relations through the exercise of power in political struggle.

Orlando Fals Borda firmly plants the roots of PAR in the intellectual and ethical challenge posed to traditional social science practices by anticolonialist activists in the global South, and others have situated it in the industrial democracy movement in Scandinavia,[6] but important contributions to PAR have been made by social movements in the United States as well. New social movements have often used "people's science" and participatory research approaches, including the Black Panthers' health education and research programs, the women's health movement, and AIDS treatment activism.[7]

PAR emphasizes the link between political structure and knowledge production, starting from the lived experience of ordinary people and undertaking collaborative research design, data collection, and analysis.[8] The approach's explicitly participatory nature and action orientation enhance the reliability and validity of research results because (1) PAR integrates the points of view of multiple analysts, (2) its action orientation tends to check what people say in interviews against what they actually do in practice, and (3) the action undertaken in the course of the research results in lasting social outcomes—infrastructure and social benefits that accrue to the participants in the research—which continue to test the results of research in the real world of community.

Participatory Design

Participatory design (PD) is an approach toward computer systems or industrial design in which "the people destined to *use* the system play a critical role in *designing* it" (Schuler and Namioka 1993, xi). PD grew out of democratic workers' movements in Scandinavia, where the high degree

of unionization and legislation that requires worker discussion of technological change in the workplace has laid the foundation for the considerable support participatory practices have enjoyed there (Greenbaum 1993,
35).[9] Like popular education and PAR, PD places a premium on the active
involvement of people who most directly confront problems (Levinger
1998). It draws on a broad understanding of technological and organizational systems as networks of practices, people, and objects embedded in
particular contexts. Though PD tends to focus narrowly on information
system design, there are many ways that a PD methodology can be used
to strengthen community-building efforts and critical citizenship projects.
For example, the feminist literature on urban planning, which combines
the methodological focus of the field with feminist critical analysis of the
gendered structuring of social space,[10] broadens the domain of PD to
provide a rich source for practicing participatory design outside the workplace. Feminist writers describe strategies that engage stakeholders in technological development, broadly construed: neighborhood revitalization,
economic development, and dweller-controlled public housing.

Popular technology draws on the best lessons of these three traditions
to identify, analyze, and articulate the relationship among people, politics,
and technology. Unlike access-oriented approaches to high-tech equity,
popular technology does not seek to help poor and working-class women
"reintegrate" into the information economy, which they already support
through unpaid and low-wage labor of many kinds, but rather seeks to find
ways to reform the information age so that it can support fuller humanity
for all people.

Popular Technology Projects

WYMSM, with support and input from the YWCA and local community
members, undertook nearly a dozen ambitious popular technology projects
between January 2001 and September 2003. These projects were designed
to make visible the role information technology plays in economic and
social justice concerns in a region increasingly dedicating itself to high-
tech growth. Owing to space considerations, I am unable to describe all of
these projects, though I summarize them in appendix D. In this chapter I
describe three projects in detail: the construction of the YWCA community
technology lab, the development of the online Women's Resource Directory, and the design of a popular education exercise called Beat the System:
Surviving Welfare. These projects took place at different times and under
vastly different circumstances, but I believe they show a progression from

Box 6.2
WYMSM Member Profile, Jennifer Rose

"Don't be scared to stand up for what you believe, even though it may have costs."

I've always really loved computers, since high school. I didn't own one until I was in my early twenties, because I couldn't afford one, but I've always been interested. I thought they were the way to go in the world. When the dot-com bubble burst, a lot of that stuff was going away. But then game design was getting big. So I thought to myself, "Maybe I can make something more than these stupid shoot-'em-up games. Something that might actually make people think." That's why I got involved in WYMSM. I had come to the YWCA to escape a domestic violence situation, and when the project was mentioned as being a possible computer game, that sparked the geek side of me.

Before WYMSM, I was very apolitical. I wanted NOTHING to do with politics. Politics? Yuck. It was like, "Leave me alone, I'll leave you alone." But WYMSM helped me realize that I wasn't the only one going through this stuff. I thought, "Hmmm. Maybe if we band together, we can actually do something." That's why my favorite WYMSM projects were the Pavilion event and the Hunger Awareness Day simulation. We heard what people in the community thought, and we also found out what they *didn't* realize. It opened up a few more eyes. Now, [people who attended these events] might think, "What resources do I have with state policymakers? Can I have my voice at this town meeting, or these city council meetings, or fight against cuts in funding for that program? Maybe my voice might make a difference!"

In WYMSM, it was neat working with people from a variety of different backgrounds, but there were so many similarities among us, too. I started to think, OK, this is a universal thing. What can we do about it? WYMSM made me realize that it was time for me to step up and state my political views, even though they might not be the popular ones.

Five years from now, I hope to be working on a degree in library science. Unfortunately, I still have a lot of loans to pay off, and there have been a lot of cuts in education where I live now. It's getting harder and harder to get into schools, even community colleges. Even if you get in, it's almost impossible to get into some of the classes. The governor is cutting things from the people who can least afford to have things cut. Funding for domestic violence. Homelessness prevention. Health care, mental health. It's getting BAD.

I'm doing well now, but it took a while—I'm OK because there really are a lot of amazing programs here. I went through a number of them trying to get safe and on my feet. I would like the world to change so that women and children—actually, all people—are more empowered to know what's out there, know what they can learn, and have access to broader horizons. Maybe if they hear other people's stories, maybe they won't think they are alone.

Based on a phone conversation that took place on October 6, 2009.

fairly straightforward projects centered on outreach, technology training, and community service to more transformative techniques of participatory research, critical inquiry and social change.

The Community Technology Laboratory

Hoping to build on the public interest generated by the Hunger Awareness Day event, I collaborated with Pat Dinkelaker and Christine Nealon to write a grant proposal to the city of Troy for the design of an asset-based community development tool. The grant was intended to provide neighborhood residents with tools for mapping the community's assets and sharing that knowledge through a database-backed Web site.[11] Though we did not receive the funding, several months later we received a letter from the mayor of Troy offering to donate eight personal computers for the creation of a community technology laboratory at the YWCA.

From the beginning, the community technology lab was seen by the YWCA staff as a way to realize the Sally Catlin Resource Center's mission to provide a place for women to gather and "find the tools they need to craft the lives they want for themselves" rather than as a site for simple technological training. Participatory planning and design of the lab was therefore crucial. In addition to informal conversations with YWCA community members, I held four participatory planning sessions, called "Tech Lab Coffee Talks": two in January 2003, six weeks prior to the lab opening, and two in May 2003, three months after we opened the doors.

In the first two sessions, participants offered input into what kinds of classes the lab should offer and made suggestions about dealing with operating issues. Advice from these community workshops included the following:

• Develop a community technology advisory board (made up of residents and other community members) that can make suggestions for the growth and maintenance of computer programs at the YWCA.
• Involve as many people in the lab as possible by holding an Open House period when the lab opens in February and scheduling lab orientations during that week.
• Help establish strong knowledge of computers by offering as many classes as possible, though participants agreed that having "open studio" and "group project" time was also important.
• Create a gallery of work that people have done or are doing in the lab, so that the community can share its talents and expertise, and so everyone can see what kinds of things are possible using computers.

Figure 6.1
The YWCA community technology laboratory. Clockwise from bottom left: WYMSM members Jenn Rose, Julia Soto Lebentritt, and Jes Constantine.
Photo: Virginia Eubanks

Figure 6.2
"Welcome to the YWCA Community Technology Lab."
Photo: Virginia Eubanks

• Offer tutorials and courses online as much as possible, so that people can take best advantage of open studio time and continue to craft the technological education they want for themselves.
• Provide as much information as possible about the YWCA's administrative challenges (such as lack of money for supplies and operating expenses) so that everyone in the community is on the same page about possibilities and limitations.
• Draw on the community to provide hostesses for the lab so that we can provide for our own needs internally whenever possible.

In response to these suggestions, I developed a schedule that alternated between open studio time with a lab hostess available to offer help when needed; group project time, during which participants could engage in loosely structured projects; and formal classes taught by YWCA staff, hostesses, or interns and volunteers from local universities and schools.

Christine Nealon and I also developed a tech lab user agreement, which made organizational constraints—particularly monetary restraints—clear, asked for advice in overcoming them, and communicated the possibilities of the lab to YWCA community members. We designed a series of formal short-term courses, including courses in the Microsoft Office Suite, a "Great Women of Troy" research and desktop publishing course, computer basics, a résumé workshop, Internet basics, and a computer use and maintenance course called "How that D@mn Box Works."

Traditional technological training classes like these, however, were only intermittently attended. The longest-running—and most successful—activity in the lab was Audio Group, an informal weekly meeting of women from both inside and outside the walls of the YW who gathered to write and play music, edit audio on the computers, and produce projects like "Sewing for Survival," a set of oral histories about women in the Troy area who historically used sewing as a means of attaining self-sufficiency and exerting social solidarity and power.

At community meetings held three months after the opening of the lab, one community member said she thought the tech lab was the best thing to happen to the YWCA that year. Some community members, however, felt that their access to the lab was restricted and that programs lacked consistency. This was in part a result of organizational constraints: the YWCA did not have an operating budget that allowed them to pay a lab hostess, and existing employees already had their plates full. Participants in the second round of coffee talks identified a potential solution to the YWCA's resource constraints when they pointed out that hostessing is the real work of the lab, and that hostessing was something to which many YWCA residents were perfectly suited.

If we want people to come back, they argued, our focus should not just be getting the door open. They suggested paying YWCA residents rent credit for gathering people in, checking in with them as they leave, and making personal connections. When we instituted this change, though it offered some challenges with administration and oversight, the schedule quickly filled with facilitated lab time (Figure 6.3). The lab opened on February 12, 2003, and operated continuously until an abrupt change in leadership at the YWCA in July 2005 shifted the organization's guiding framework from a social justice model, which saw residents as resourceful members of a social movement, able to provide for the organization's needs internally, to a service provision model, which sees residents as women in need of assistance, skills training, and self-esteem building.[12]

Several lessons about building community technology capacity can be drawn from this example. The first is that grant writing is an important

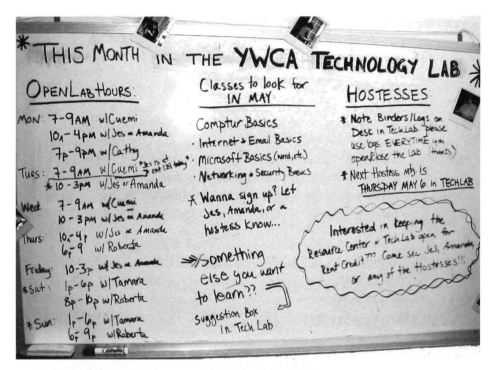

Figure 6.3
YWCA community technology lab schedule, April 2003.
Photo: Virginia Eubanks

role that engaged academics can play in helping overtaxed organizations access resources. But I should not overestimate this strategy's efficacy: the YWCA did not receive the grant that I wrote, though it did receive eight workstations based on a personal communication and a brief letter written by Executive Director Pat Dinkelaker. The second lesson is that, sometimes, taking on a more modest role in these projects results in greater success. Often, scholars' focus on stated initial outcomes, required by funding entities to be stated as "deliverables," stands in the way of an organization using resources to suit its own strengths and needs. In some sense, my involvement with the lab, focused as I was on creating technology training and job skills classes, had a cooling effect on residents' use of the resource. Despite the participatory process of the lab design, my framework for understanding what people should be doing in the lab kept them from undertaking activities that worked for *their* lives. This was corrected only when community members suggested paying rent credit to YWCA residents to act as lab hostesses, which increased lab use dramatically. Finally, the technology lab illustrates the need to connect technology to nontechnology, mission-based goals identified by the community itself.

Despite these cautions, it is fair to say that the community technology lab was a success. It built capacity at the YWCA, and the lab served as an interface between the YWCA and other organizations. Soon after it opened, for example, the Hudson-Mohawk IndyMedia Center moved into another room on the floor and began offering media and videography training to women in the YWCA community. The lab also fostered the YWCA ethic of mutual aid. Hostessing is a peer education role, and the lab drew on the skills and resources of the community to develop new skills and access outside resources. This peer-based process resulted in a final indicator of program success: a multiplier effect created by the lab's focus on developing internal resources and leadership. The lab provided participants with the resources to reach out to become teachers in the larger community, broadening social networks and expanding priorities beyond operational viability.

During the Great Depression, Myles Horton wrote, "The spreading of knowledge through a large group is a difficult task, but the widespread understanding of the social forces which will enable the masses to assume social control is essential to the welfare of our country. This means that adult education must be cultural rather than vocational. The disinherited worker must be awakened regarding his destiny; not trained to do better work for his industrial masters" (Horton 1933, 3). Building technical capacity and providing vocational or skill-based training is important. But it must be integrated into a total educational process. If this integration is successful, it produces not only skills but also feelings of hope and competence and an ability to identify problems, articulate solutions, and collaborate with others on goals. The process of mutual education, of sharing knowledge and resources with others, makes it possible to envision a community technology lab as a place for social movement, a site for the development of critical technological citizenship.

The Women's Resource Directory

A grant from the Verizon Foundation made it possible for Pat Dinkelaker and a local computer programmer, John Fudjack, to begin developing a demonstration project, the online Women's Resource Directory, in 2001. Central to the project was a database, designed by Fudjack and accessible to the public and service providers on the Web, that would list programs and resources available to women in the Capital District. A "Talk Back" section would provide a place for consumers to record their experiences with local agencies. An "Advocates Online" forum would provide a listserv where advocates and staff of community-based agencies could ask questions, problem solve, and share information about programs and resources. Finally, an Internet learning laboratory would provide hands-on training for women in the YWCA community to learn how to develop and maintain the Women's Resource Directory database, Web site, listservs and other interactive services.

WYMSM members became involved in the project in December 2002, when the YWCA hired me to move the Resource Directory forward and develop the Internet learning laboratory training program. The Resource Directory offered WYMSM and other YWCA community members opportunities to develop technology skills—including data entry, beta testing, technology mentorship, graphic design, databases, and systems administration—in a paid, supportive environment. Crucial to the project from the beginning was peer learning. Participants would step through a process of learning skills and then become peer "skill sharers," who would pass those skills on to the next generation of participants.

After I spent a few weeks designing and documenting a data entry process and holding YWCA community meetings to raise interest and awareness about the Resource Directory and solicit input about the process, we hired a first round of four Resource Directory team members—WYMSM members Cosandra, Cuemi, and Zianaveva and YWCA resident Juliet—in mid-December. As a team, the five of us contacted local resource providers in the Capital District, gathered information on their programs and services, compiled and edited the information, and entered it into the online database.

In January we held a team meeting to check on our process and do an HTML training workshop. At that meeting we also developed a work agreement to structure how current team members could become skill sharers. We held two more community meetings, after which we hired four additional team members, bringing our total to nine. Each new woman was paired with a skill sharer, and together the women logged 269 hours of work on the Resource Directory, collecting, editing, and inputting information on the offerings of more than eighty local social service agencies. By March 2003 I had begun turning my attention to full-time dissertation writing, and helped the YWCA find a replacement for me in the Resource Directory process. My replacement in the project was a wonderful local artist and activist who facilitated a wealth of new programs, including the popular weekly audio group and the "Sewing for Survival" oral history

Individualized Pathways to Building Technology Skills: Resource Directory

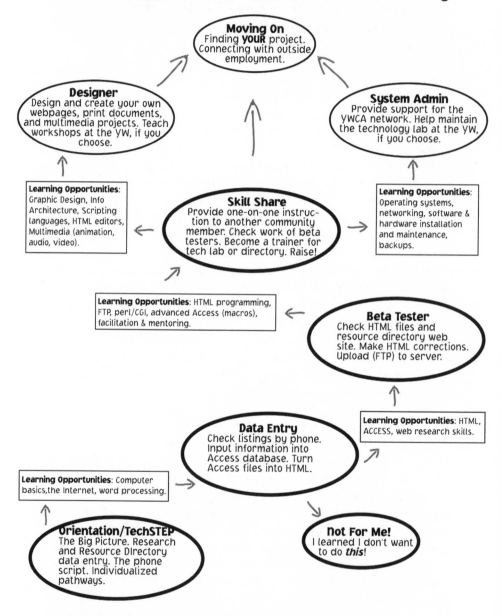

Figure 6.4

Individualized pathways to building technology skills.

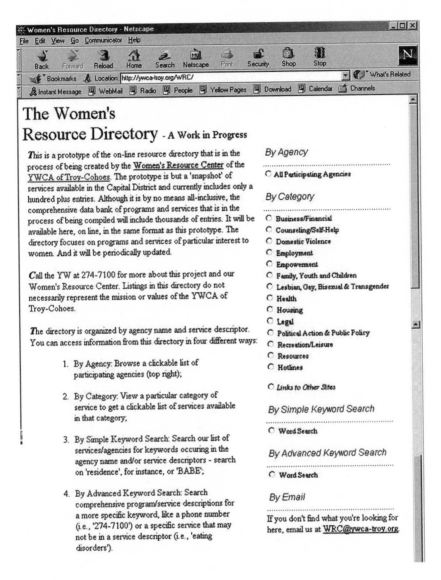

The following text appears within the browser window shown in the figure:

The Women's
Resource Directory - A Work in Progress

This is a prototype of the on-line resource directory that is in the process of being created by the Women's Resource Center of the YWCA of Troy-Cohoes. The prototype is but a 'snapshot' of services available in the Capital District and currently includes only a hundred plus entries. Although it is by no means all-inclusive, the comprehensive data bank of programs and services that is in the process of being compiled will include thousands of entries. It will be available here, on line, in the same format as this prototype. The directory focuses on programs and services of particular interest to women. And it will be periodically updated.

Call the YW at 274-7100 for more about this project and our Women's Resource Center. Listings in this directory do not necessarily represent the mission or values of the YWCA of Troy-Cohoes.

The directory is organized by agency name and service descriptor. You can access information from this directory in four different ways:

1. By Agency: Browse a clickable list of participating agencies (top right);

2. By Category: View a particular category of service to get a clickable list of services available in that category;

3. By Simple Keyword Search: Search our list of services/agencies for keywords occuring in the agency name and/or service descriptors - search on 'residence', for instance, or 'BABE';

4. By Advanced Keyword Search: Search comprehensive program/service descriptions for a more specific keyword, like a phone number (i.e., '274-7100') or a specific service that may not be in a service descriptor (i.e., 'eating disorders').

By Agency

   ○ All Participating Agencies

By Category

○ Business/Financial
○ Counseling/Self-Help
○ Domestic Violence
○ Employment
○ Empowerment
○ Family, Youth and Children
○ Lesbian, Gay, Bisexual & Transgender
○ Health
○ Housing
○ Legal
○ Political Action & Public Policy
○ Recreation/Leisure
○ Resources
○ Hotlines

○ Links to Other Sites

By Simple Keyword Search

○ Word Search

By Advanced Keyword Search

○ Word Search

By Email

If you don't find what you're looking for here, email us at WRC@ywca-troy.org.

Figure 6.5
The YWCA Women's Resource Directory.

project, in the technology lab. I continued to work on the Resource Directory off and on, time permitting, and we completed inputting information in June 2003.

That summer, we "soft launched" the Resource Directory, making it available to the public on the Web. However, because it wasn't quite perfect, we never promoted or advertised it, and it received only minimal use. Though my replacement brought important and unique skills to the technology lab, the transition was difficult for the Resource Directory project to survive, and the YWCA, pulled in multiple directions by other needs of its residents and lacking sufficient funds to truly develop the interactive components of the project, turned its attention elsewhere. We had great success with the peer-to-peer learning aspects of the project and completed the directory itself, except for some minor editing work, but the "Talk Back" and the "Advocates Online" functionality—in some ways the most interesting, and most challenging aspects of the project—were never completed.[13]

Our failure to complete all phases of the Resource Directory project illustrates a basic contradiction of trying to do participatory, movement-centered work in the context of a grant and foundation culture that mainly supports social service provision and technology access. In many ways, the participatory orientation of the Resource Directory project and the mission of the YWCA of Troy-Cohoes conflicted with the dictates of the funding agency. While the YWCA was focused on broad-based participation in designing and implementing technology programs, the funding source focused on delivery of products and outcomes, narrowly defined as the number of women moving through technology training into employment. While the YWCA focused on the social supports necessary for women to truly engage in education and learning—on the whole person—the funding source did not see the health, housing, child-care, nutritional, or emotional needs of women in the YWCA community as relevant to the program. Finally, the YWCA was interested in being involved with the project for the long haul, but the funding source was on a two-year funding cycle for demonstration projects, and there was no way to sustain the ongoing requirements of the Resource Directory, which included updating the database, managing mentor relationships, identifying possible employers, and keeping abreast of new technological innovations and community content.

The characteristics of the project that made it most successful for women living in the YWCA community—its focus on process, participatory development and implementation, acknowledgment of the whole person, and a long view of community change—doomed it to only partial success in fulfilling the mandates of a funding agency focused on vocational training for high-tech employment. But it was these very characteristics that also made the final product so innovative. Years before Angie's List, WYMSM and its allies produced a way for local women to find, evaluate, and share information about the community services available to them. It wasn't until six years later that a comprehensive resource directory became available in the area, United Way's 2-1-1 project. While 2-1-1 is an important resource, it is significantly different from the Women's Resource Directory as we imagined it—it is available only over the phone during regular business hours, its design and content lack the input of program clients, and it provides no mechanism for feedback or evaluation of the resources or programs it lists.

Despite our failure to deliver the final product we envisioned, the Resource Directory had important impacts on the women in the YWCA community who were involved in the project. First, YWCA residents spoke as equals to social service agencies that many of them had accessed as clients in the past, an experience that so resonated with them that Cosandra Jennings mentioned it as her favorite moment in all of WYMSM's history when we were writing her profile for this book six years later. Second, we successfully created a peer-based technology mentoring program. This model then became the standard for work in the community technology lab, where most project members went on to serve as lab hostesses. Finally, a great deal of relationship building and discussion about the impact of technology on women's everyday lives went on in the lab during production of the Resource Directory. The environment created by peer-learning, the value placed on participants' knowledge and life experience, and the open and transparent nature of our process produced many of the formative insights upon which this book is based.

Beat the System: Surviving Welfare

Beat the System: Surviving Welfare began in WYMSM's imagination as a suite of simulation software, not unlike EA's popular computer game, The Sims, that would both provide tools for navigating the social service system and teach a middle-class public about the unique challenges of surviving and thriving on public assistance. In a software design brief written in April 2002, we described the goal of the project:

This technological tool will help people in the community make informed decisions about their lives *now* by:

1. Building a collective consciousness about important concerns in their lives, including livable wages, existing resources and alternatives to them, personal and community capabilities, etc.;

2. Using that knowledge to make empowered decisions about navigating the existing [social service] system; and

3. Providing opportunities to take concrete action (by joining social movements) to change that system.

The software was to contain three sections. The first was a decision-making tool intended to address people's immediate needs and help them make informed choices about their near-term future. We hoped to create an upstate version of the New York City self-sufficiency calculator, an online tool developed by the Women's Center for Education and Career Advancement (WCECA) to help women calculate their eligibility for public assistance and the wage they need to earn to be truly self-sufficient.[14] The second part of the software suite was intended to be a narrative simulation, a model of triumphing in life, based on information and stories drawn from the experiences of women in the YWCA community. The third goal was to create an asset-sharing database, the Community Asset Bank (CAB), that would be used to inventory and facilitate the sharing of skills and resources present in the YWCA community.

WYMSM was emphatic from early in the design process that we wanted to create an educational resource that could communicate group members' personal knowledge of the social service system to a middle-class public that seemed quite naive and misinformed about real life. At the same time, we were concerned that the diversity of the women in the YWCA community should be represented, and that the game could be a resource to both those who had experienced the social service system and those who had not. The group decided to meet the challenge of representing a diverse community's viewpoints and experiences by doing our own research based on a "reconstructed stories" compositing method used by Irena Papadopoulos, Karen Scanlon, and Shelley Lees in their work with visually impaired adults.[15]

WYMSM interviewed six YWCA staff members about women's common challenges navigating the social service system and their strategies for triumphing despite an often confusing and irrational bureaucracy.[16] Group members then transcribed and coded the interviews with the help of a data analysis handout (figure 6.6), searching out keywords and concepts that were common across these narratives (figure 6.7). A number of themes emerged from our data, including red tape, work, child care, education, health, access to information, resources, emotional response, systematic inequality, and "world turned upside down," a phrase WYMSM members used to denote how the stories that circulate about the social service system

Interpreting Interviews   WYMSM Data Worksheet

We need to read interviews on a lot of levels, for both visibile (the factual answers to questions the interviewer asked) and hidden (patterns of responses, common or incongruent understandings) content. It may be helpful to develop a list of keywords and/or concepts for each interview and to mark on the interview where each came up (this is called *coding* in the social sciences).

Keywords	Concepts

You might also consider (and answer) the following questions while reading interview responses:

1. What suprises you the most in the interviews?

2. What is most interesting? Most useful?

3. Where do interviewees agree? Disagree?

4. What questions elicited "the party line," i.e. answers that are suspiciously similar? What does that mean?

5. What would you like to ask (that you didn't), now that you've gotten a chance to think about it a little?

6. What question(s) didn't work?

7. What aren't they saying?

Figure 6.6
A worksheet on interpreting interview data.

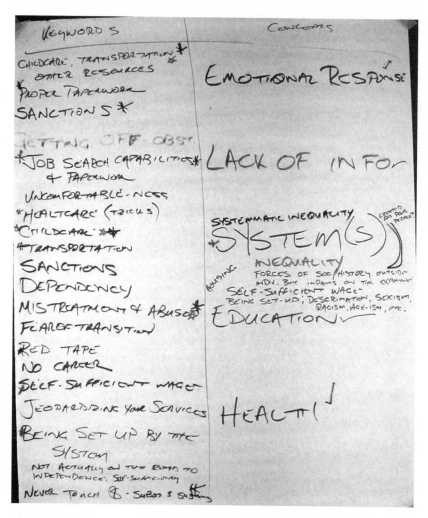

Figure 6.7
Flip chart developing concepts from "keywords" in the interview data.

(and its users) in popular media and policy do not match women's direct experiences.

WYMSM member and author Julia Soto Lebentritt then led the group in constructing composite stories based on the characters, keywords, and themes developed from the data analysis.

In box 6.3, for example, is the "Andrew" composite:

Box 6.3

Andrew

> Andrew is a thirty-three-year-old intravenous drug user recovering in a MICA program. He has hepatitis C. Although he receives Medicaid, the federal health insurance program for indigent people, his bills exceed what Medicaid pays for the expensive pharmaceutical drugs that are essential to his liver functioning. His drug bills run about $4,000 annually. Seven months into the year, his Medicaid coverage runs out, and he is denied for the rest of the year. He has to switch to one primary care physician who works for a clinic that will accept just what Medicaid pays for and not bill him for the balance, and drop all his other specialists.

Finally, the group intended to take these reconstructed stories to a next level of interviews. WYMSM members would go over the composite stories with other YWCA residents and ask the following questions:

- What are your general impressions of the stories?
- Do you identify with the issues in the stories?
- Are the stories true and credible to you?
- Are any of the issues in the stories strange to you?
- Do you disagree with any of the stories?
- Are any stories missing?

Before the group could conduct a new round of interviews, however, we were invited by local activists Andrew Lynn and Anne Marie Lanesey to develop a workshop for a public daylong skill-sharing event called Pavilion. The group used its next few meetings to create a popular education exercise using the composites as the basis for skits to be performed and discussed by community members attending the event (I have provided the exercise in appendix C, including the additional composite stories we developed).

New insights were generated by public participation in Beat the System skits. Workshop participants explored possibilities that had not occurred

Figure 6.8
Poster for Pavilion event.

to WYMSM members or YWCA staff members. For example, in telling Andrew's story, workshop participants portrayed a number of scenarios by which he could attain the health care he desperately needed for the five months a year Medicaid fails to cover. One possibility was that Andrew could become a test subject for experimental hepatitis drugs, but the group's final solution to his dilemma was to have Andrew commit a crime brutal enough to get five months in county jail, where he would get the medical care he needed, but not one so severe that he would have to go to federal prison. Through this exercise, the public—even those with no experience with social services—seemed to understand the double binds of public assistance and the potentially catastrophic "choices" many clients are forced to make. There was even some humor. When the community member playing Andrew committed his simulated crime, he mimed stabbing another audience member—local Green Party leader Mark Dunlea—at which point an audience member cried out, "Not the Green Party!!" The interplay between teachers and learners, the humor, and the uncommon insight yielded by the process are typical of popular education.

WYMSM was drawn to compositing because the data we gathered were very personal and we were initially uncomfortable asking interviewees direct questions about their experiences with public assistance. Composite stories offered respondents a way to depersonalize their own information by identifying (or not) with semifictional characters. The "composite and response" process allowed us to interrupt the extractive and demeaning interview experiences many community members have endured in the social service system. By performing, coding, and analyzing interviews, group members came to understand their own experiences and those of others at a deeper and more personal level at the same time that we built from group experiences to create policy parables with national—and even international—reach and scope.[17]

Creating More Critical Technological Citizens

At first glance, it may seem that the *technology* part of popular technology disappeared over the course of these three projects. The community technology lab is clearly a technological project in the most literal sense of the word, but it can be more difficult to understand the Beat the System exercise as popular technology. But the move from direct vocational training to critical technological education, which creates a context in which a different form of engagement with technology can emerge, is exactly the point of popular technology. Our interest in technology did not disappear;

technology did not evaporate. Rather, our understanding of technology expanded. Over time, rather than focusing solely on making technological products, WYMSM created collective practices for analyzing and intervening in the interlocking issues that make up a high-tech equity agenda. These insights upended traditional understandings of the relationship between technology and women's poverty in the United States and resulted in ongoing projects that focus on the hidden inequalities of the information economy, the relationship between technology and state violence, and the role that information and communication play in supporting or undercutting human rights.

WYMSM's multiple areas of organizing—high-tech equity, poverty reduction, and welfare rights—are not conceptually separable in the real world of information technology. Too often, science and technology policy scholars see the domains of IT and social policy as distinct, denying that studying economic, political, and social inequality is integral to the social studies of science and technology. But "technical" policy issues, such as the pace of change, the size of sociotechnical systems, and the nature and governance of innovation, and "social" policy issues, such as poverty, racial inequality, and gender discrimination, interpenetrate at every level. The same political and economic forces shape both poverty policy and IT policy,[18] and all policymaking, including IT policy, inevitably includes social politics, especially the politics of class, race, and gender.

If the goal is to foster critical technological citizenship, we must see high-tech equity issues in their social, political, and economic context. This is true from an ethical standpoint as well as from an empirical standpoint. WYMSM's refusal to see poverty, welfare rights, and information policy as conceptually separable led to new insights that would not have been available if we had stayed in one "issue silo." Our holistic viewpoint revealed, for example, that women's relationship to the social service system is deeply mediated by IT, and made women's presence in the low-wage, high-tech economy visible. Technology, inequality, and citizenship are deeply linked in the lives of women in the YWCA community. Popular technology, as experiential education rooted in social justice, refuses to separate lived experience into conveniently abstract categories or to isolate high-tech equity issues from their real-world context.

Popular technology differentiates "training" and "access" from "critical citizenship" and "justice." By focusing on creating more critical technological citizens rather than producing more adept technology users, popular technology projects such as the community technology laboratory, the Women's Resource Directory, and the Beat the System exercise build on the emphasis on critical literacy and social change central to popular education, participatory action research, and participatory design.[19] Differentiating critical technological citizenship from "training for integration into the information economy" was crucial for developing a strong relationship with women in the YWCA community, as some participants felt forced by the larger society to become technologically literate. As Zianaveva Raitano explained, "When society mandates that you have to use technology—they want society to be constructed around it for whatever reason—then [to] adults that are already established and in adulthood without that . . . it seems like a chore." Popular technology moved to counter the extractive role that technology plays in the lives of women in the YWCA community by drawing on their knowledge, building a space where that knowledge can be analyzed collectively, and creating opportunities for broader political engagement.

Community technology programs that focus only on developing technical skills and proficiencies have low coherence and provide few resources for developing critical technological citizenship. WYMSM found that the real work of popular technology was contextualizing everyday experiences in the information age through collaborative research and education projects. Skill building and participatory design of technological tools were important but secondary by-products of the far more important process of collective personal and political transformation. Popular technology was less about making things or teaching skills, and more about exploring ways of knowing, creating spaces for the collaborative analysis of our shared technological present. The central challenge of popular technology is, thus, an epistemic challenge. Popular technology's model of justice is cognitive, not distributive.

7 Cognitive Justice and Critical Technological Citizenship

Epistemology is politics.
—Shiv Visvanathan, "Knowledge, Justice and Democracy" (2005, 84)

Accounts of a "real" world do not, then, depend on a logic of "discovery," but on a power-charged social relation of "conversation."
—Donna Haraway, *Simians, Cyborgs and Women: The Reinvention of Nature* (1991, 198)

Many scholars and policymakers view participatory democracy as nice in theory but implausible in practice. But broad-based and meaningful participation in the decisions that affect our lives creates important overall benefits, both for individual participants and for democratic society (Polletta 2002). Participatory decision making creates solidarity, provides a primordial soup of diverse ideas and experiences from which innovation arises, teaches participants about politics by engaging them in politics, sustains commitments to social change, and provides personal, social, and material support for those commitments. WYMSM provided the base from which all other projects—the Hunger Awareness Day Event, the Women's Economic Empowerment Series, the technology lab, and the Resource Directory—developed and grew. The group's "endless meetings" were not conceived specifically as outcomes or deliverables when we began the project, but in retrospect, group members agreed it was the meetings—the camaraderie, attention to process, willingness to share, and fellowship— that provided the most important, meaningful, and lasting impacts of our collective work.

What lessons about the relationship among technology, citizenship, and social justice does WYMSM's experience offer? In interviews with WYMSM members after the disbanding of the group, I found that members experienced frustration with some aspects of our group process but that

Box 7.1
WYMSM Member Profile, Jes Constantine

"I learned about the kind of woman I want to be."

I was young. I feel like I was very, very young. I was a freshman or sophomore, only nineteen when WYMSM started. I got to the YWCA through the public service internship, a course that Nancy Campbell was teaching. With a few other people, I helped create all the technology pieces of the Sally Catlin Resource Center—the wiring, infrastructure, installing the computers and networking them. We were laying the groundwork for this new dimension of the YW, this whole new way of thinking about technology.

The YWCA really connected with me. I loved being there. Just talking to the staff, hearing them talk in our meetings, but more important, seeing how they were when they *left* a meeting. I could see how they interacted with their professional peers, but when they left a meeting, they had the same amount of respect, and time, listening to the people who lived at the YW. It was a whole different way of thinking, and I wanted to be a part of it. So I worked there that summer as a human potential advocate, and then we did the Women, Simulation and Social Change workshops in the fall, and WYMSM started that winter.

WYMSM changed the way that all of us thought—the way we perceived ourselves, our relationships, our roles, our lives. There was a shift in people taking responsibility, and learning, and wanting to change things. We were taking control, asking questions we hadn't asked before. I was learning that women can be independent thinkers. I thought of women as very strong, because I was involved in sports, but I didn't really understand feminism at the time. I'm glad that my first experience of feminism was with women from all different walks of life. That was eye-opening.

WYMSM was also the first time I enjoyed a meeting. Our business was doing deeply personal things. I got to do some facilitation, and that was fun, focusing on the process. We learned to work together, to deal with conflict—and we certainly had some conflict!—and to be OK with the uncomfortable. That has carried on for me. I find myself becoming the facilitator, even when I'm not supposed to. I just slide into that role, because I learned how!

Five years from now, I want to be working with kids, preferably older kids who don't normally have access to technology education. I'm excited to go to graduate school to get a teaching certification and a master's in applied technology education. Basically, I want to be a computer teacher for a high school. Even more than teaching, I want to coach!

Popular technology ruined me. I can't just teach a computer class. I can't be like, "Here's how you make a spreadsheet." I want to help people use

Box 7.1
(continued)

technologies in ways that suit their lives and purposes. People learn better if whatever they're learning has relevance to their everyday life. I'm interested in helping people appropriate technology. I want to have conversations like, "This computer doesn't work for you because it was designed by *these* people who live *this* kind of life. How would *you* design it?" And obviously, I want to talk to people about "access." You might not think you have access, but let's think about it more, think about how you *do* interact with technology.

I was always learning at the YWCA, and I want readers to know that you can always learn in places you might least expect to learn, and from people you might least expect to learn from. I feel like I learned about life, I learned how to be a woman. I learned about the kind of woman I want to be.

Based on a conversation over lunch, April 29, 2009.

the message sent by the group's democratic decision-making processes, the diversity of the group's membership, and the catalyzing effect of WYMSM on other political projects had lasting impacts on members' beliefs about citizenship and their feeling of entitlement to political articulation.

Transforming Technological Citizenship

There is not a great deal of academic work, either in the political sciences or in social studies of science and technology, about the relationship between technology and citizenship.[1] Yet our political identities are bound up with technological artifacts and processes in complex and important ways. Technologies of everyday life, embedded in political, educational, bureaucratic, and social settings, actively shape our identities, our communities, our institutions, and our relationships. They affect how we relate to each other and how we understand ourselves. They teach us lessons about who we are, and shape our political and cultural voices. They help distribute material and informational goods, but they also structure our social and political imaginaries, our sense of what is possible, acceptable, and just.

Think for a moment about the ubiquity of technology in your life. Do you have a TV? A computer? A newspaper? Do you drive a car? Ride a bike? Take the bus? Do you have a home phone? A cell phone? Struggle to find working pay phones? Deeper still, are you affected by the media? By the interstate highway system and your neighborhood streets? By the global system of cell towers, telephone lines, and satellites? Questions about the shape and impacts of these wide-ranging technological systems are too often answered by a small group of people—engineers, scientists, regulatory agents—who have a necessarily limited understanding of the interaction between technology and everyday life. In a democratic society, we *all* should have a right to make decisions about how these broad-ranging, systemic forces affect our lives. To participate more fully in democratic life, all of us must be more critical, more thoughtful, about these relationships.

When technology is broadly understood as a shaper of society and political subjectivity, it becomes clear that no one group of individuals has a monopoly on technological expertise. It is useful to have information about how a specific technological artifact might operate in a given setting, or how the nuts and bolts of some particular system work. But this kind of specialist knowledge is not the only kind of knowledge necessary for making decisions about how to live more fully, and more fairly, in our shared technological present. As Brian Wynne has written, you don't need to know how to work a nuclear reactor to decide whether or not you want one sited in your community (Wynne 1996). This is the goal of popular technology: to create a space in which a wide variety of people can collectively consider our shared technological present, to open up a conversation about how to be more critical technological citizens.

Transforming technological citizenship begins when people who are marginalized in technoscientific and economic debates claim their right to political articulation and are seen as epistemological agents able to define and solve their problems for themselves. As Shiv Visvanathan argues in his 2005 essay, "Knowledge, Justice and Democracy," efforts to truly democratize science and technology should focus on epistemology rather than on information delivery, on democratizing critiques of science and technology rather than on public understanding of science or the consumption of science and technology products. There have been exciting developments in this direction across the globe. New critiques in India from social movements such as the Kerala Fishers Forum and the Narmada Bachao Andolan (Save Narmada Movement), for example, focus on how the impact of technology on the economy, government, and public sphere offers opportunities to renegotiate the social contract (Visvanathan 2005, 86–89). But in the United States, social movement actors and the makers of public policy have largely failed to escape the stranglehold of diffusionist, distributive approaches to high-tech equity to address these broader technopolitical issues. We think too little about citizenship, too little about the kind of social contract we want for the information age.

Technologies of Political Articulation

Initially, many women in the YWCA community had an ambivalent relationship to politics. For example, in a WYMSM meeting in March 2002, we did a team-building activity called "colorful cards." Each person was given a set of three or four cards with statements printed on them and had to form affinity groups by interacting with other members to find out who else identified with the statements on their cards. Jes Constantine, explaining the activity, used the example of a card that was printed with the phrase, "I have been involved with politics." Not one of the nine participants in the activity (except me) identified with the statement. When I expressed surprise that no one in a group that calls itself Women at the YWCA Making *Social Movement* would say she had been involved in politics, group members explained that they interpreted the phrase to mean, "I am a politician," "I have held elected office," or "I have been involved with a political campaign."

Many WYMSM members, and other women in the YWCA community, began popular technology projects with a deep, abiding distrust of politicians and the political process. This distrust stemmed from two main sources. First, many women in the YWCA community interact with public institutions such as the social service system in ways that leave them hypervisible to scrutiny, stigmatized, and deeply vulnerable, both economically and physically. Second, many women in the YWCA community have a strong sense that the traditional political process does not represent the interests of poor and working-class people, believe that politicians are more interested in power than in people, and hold misgivings about building coalitions with a broader public that rarely recognizes their unique knowledge, skills, and life experience.[2] This distrust certainly could contribute to WYMSM members' hesitancy in claiming a conventional political identity or in seeing their activist and community-building work as political work.

When I first met Cuemi Gibson in October 2002, for example, she was attempting to declare herself a sovereign citizen because, as she later explained, "I don't think [African Americans] should pay taxes in America. Basically, my feeling is: you're cutting education, women don't have money for [utilities], they're sitting there in the dark, no Section 8, no federal housing. I'm at the point where I don't want to pay taxes. Why should I? I haven't agreed to anything these people are doing. I know I didn't vote anyone into this world to cut education!" Similarly, Dorothy Allen, when asked if she considered herself political, responded,

I wouldn't say political. I would say opinionated. If I had to choose a party, I would be a Democrat, but all that shit is crap to me. They're all liars, they all say stuff because they want something, and they need the people to get that something. . . . I'm more interested in thinking about homeless people, or thinking about why that lady's walking down the street—last week she wasn't smoking crack, this week she's smoking crack. Something happened in her head . . . stuff like that. I could give a damn about politics!

Cosandra Jennings argued that poor and working-class people cannot count on partisan politics to represent their interests, an opinion I commonly heard at the YWCA:

Cosandra: [T]hem giving even an inch is impossible. . . . They're rich, why would they share? The only way that could ever happen is through politics. To come together. But with the Democrats, we know that ain't going to happen!
Virginia: Yeah—even the Democrats aren't pro-poor, are they?
Cosandra: No. But they're better. [In WYMSM], it was very interesting to hear other opinions from people—from people that had drug problems to drug dealers, people on social services—it was very interesting. [But] it kind of made me sad in one way.
Virginia: How so?
Cosandra: Because of how our society is and how we deal with people that are in poverty. People that are trying to make a social change and move forward for them[selves] and their families. . . . If you are in poverty, society is against you.

Nevertheless, Cosandra and other participants found that their feelings of entitlement to political participation grew through their work with WYMSM. WYMSM members began to define politics for themselves, and claim a right to grassroots power. As Cosandra continued, "I came to terms [with the fact] that people have different problems in society, and you can look at both sides—the political—how to deal with it, and how we could help, as little guys. It encouraged me to get more into politics, to get more into society, our community." When I asked Zianaveva Raitano if she thought WYMSM was a social movement, she replied, "Yeah. Because we were talking about subjects that maybe people have written books about, but we were talking about the welfare system *directly*, and going out and interviewing people. The difference is that people that live here [in the YWCA] deal with social services. We're looking at, and questioning, social services' ability to keep people out of poverty. A revolutionary perspective! And grassroots as well. Evaluating these age-old institutions and the cycles and the systems. And we were questioning it."

When I asked Cosandra what the goal of WYMSM was, she replied, "More, better politics." To WYMSM, "more, better politics" came to be understood in epistemological terms. WYMSM demanded that the

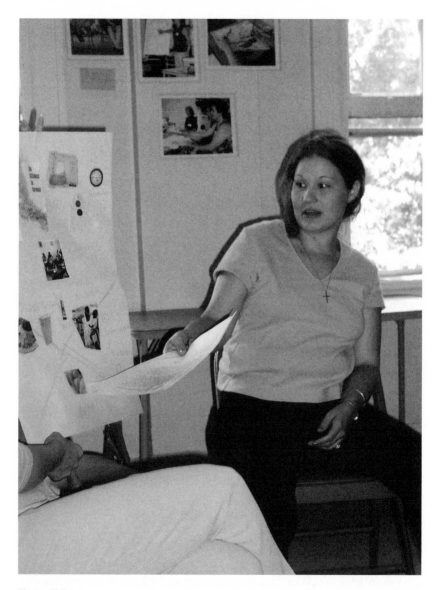

Figure 7.1
Cosandra Jennings teaching participants in the Women's Economic Empowerment
Series about global processes of production.
Photo: Pat Dinkelaker

knowledge of YWCA residents be taken seriously as legitimate expertise. Cosandra defined the political as the process of "looking at both sides"; Zianaveva defined revolutionary politics as questioning, and acting on information from, people most impacted by political institutions like the welfare system. Dorothy described her community-building efforts as centered on a will to understand structural violence—the social forces that constrain people's ability to make decisions that would have a positive impact on their lives. Cuemi eventually abandoned her quest to be declared a sovereign citizen and claimed the title of community organizer, going so far as to help develop a Get Out the Vote campaign at the YWCA during the presidential elections of 2004.

Popular technology helped women in the YWCA community develop a sense of entitlement to political articulation, provided a space for solidarity to grow, and increased participants' feelings of competence as political and technological citizens. As their desire for engagement with traditional forms of politics grew, so did their desire for engagement with a "technology for people." I asked Zianaveva, for example, how we could best support the kind of "learning between levels" she and Cosandra called for in their articulations of social movement. She replied, "One way is technology. But other ways are through projects like the [Beat the System] game. That isn't just *saying* something, that's showing realistically the challenges and presenting . . . a realistic view coming from the perspective of a poor person or a dependent person. That speaks loud." It was technological citizenship in this sense that popular technology projects—WYMSM meetings in particular—nurtured. Though WYMSM often failed to follow through on our technological "deliverables," our projects and process did much to enrich women's already nascent sense of political competence and authority.

Democracy Is an Endless Meeting (So Pay Me for My Time)

Our first WYMSM meeting took place on January 12, 2002, and the group decided to disband at a facilitated retreat on July 26, 2003. In those nineteen months, WYMSM met forty-three times, with most meetings taking place on Saturdays and lasting three hours. These meetings were the heart of our process, providing time to work on projects, a context for having difficult conversations about economics and equality, a shared sense of purpose, and significant time for less structured socializing. In September 2002, Christine Nealon described the meetings to new potential WYMSM members: "We've had kind of a nice balance between action and learning and growth and, you know, sitting around having dinner."

Box 7.2

Imagine a meeting, a group of eleven women and one man, ranging in age from nineteen to fifty-eight, two African American, one African, one Puerto Rican, eight white. We are two graduate students and two undergrads, an ex-mattress factory worker, a "lullabologist," a child-care worker, a full-time mom, a professor, a nonprofit program director, a network administrator, and a health care aide. Six of us live in the YWCA, six do not.

We meet in the Sally Catlin Resource Center, a large, sunny, open room on the second floor of the YWCA. We are surrounded by plates and cups left over from our lunch, which several members prepared and brought in, potluck style. The kids, two young boys, are playing quietly on the computers or on the rug with blocks and games. If we're less lucky, they are distractedly perched in moms' laps, coloring on our agendas and notes. On most days, we are wearing silly necklaces—red blocks, orange spheres, green pyramids—that we use to choose roles for each meeting: facilitator, fun activity leader, note taker, mission reader, and "vibes," the person who keeps track of weird feelings or interactions that may come up during meetings and reflects them back to the group for discussion.

We have a full agenda.[3] Discussion over lunch results in a full-scale "check-in," a time we set aside in each meeting to let each other know what is going on in our lives, and to collectively problem-solve major issues and crises: a sanction from the Department of Social Services, an illness that cannot be treated because of lack of health care, parenting stress, a job loss, substandard housing, family struggles, relationship tensions. Though we are already deeply in each other's business, we still do a quick ice-breaking activity to get ourselves into a meeting frame of mind: The Name Game; "I feel . . . because . . ."; Purple Panda Bear; Two Lies and a Truth; Step In; Heads Up. Many activities involve us clapping our hands, stomping our feet, or singing. The activities that get us out of our seats always work best to get the ball rolling—a collective activity that makes us all look like fools does wonders for group cohesion.

The bulk of the meeting is taken up with our current project: designing the Beat the System game, coding research data from interviews, or planning a public event. About half the time we break up that work with an activity to help us clarify our process: listing the group's assets and skills; clarifying our long-term goals; reassessing the collective process; discussing the ideal balance between process and outcomes. The other half of the time we engage in activities that help us learn something new, or connect our local efforts to the big picture: a technology training class for HTML or graphic design, a popular education exercise on the gendered division of labor, the global assembly line, or local histories of political resistance. Sometimes we take a

> research trip or a field trip. We go into the community to take pictures for
> the game, meet with other local community-building organizations, or take
> part in events or conferences. Occasionally we do fundraising activities, espe-
> cially grant writing.
>
> When needed, a WYMSM member or outside resource person leads the
> group in a training exercise: Web design, imaging, video or audio recording,
> or group facilitation. At the end of each meeting we set the agenda, the date,
> and choose roles for the next meeting. We often set goals for members to
> achieve in the two or three weeks before our next meeting. At the end of the
> day we clean up, gather the kids, hug or nod or even shake our fists at each
> other, and then, generally a little exhausted, we part until next time.

Emotion was central to our process. The passionate side of community-
based research and action—what Sandra Morgen calls "the politics of feelings" (Morgen 1983, 1995)—is rarely explored in academic writing, as
outside the feminist literature emotion is generally shunned as a force that
threatens the objectivity of a researcher. Part of the success of this project
in building space for political articulation, though, arose out of members'
willingness to be vulnerable, to weather the storms of emotional conflict,
and to be open to transformative change.

Meetings were often filled with laughter and teasing, which spilled into
trash talk over games of spades played upstairs on the halls or faded into
quiet moments of continued conversation over coffee at my apartment
across the street. There was plenty of intense conflict—some political, some
racial, some generational—though not as much as one might expect of
such a diverse group. We did not always agree, we did not always like each
other, and we did not always understand each other, though we developed
strong enough affective bonds to continue to try. Some of us became
lasting friends, but friendship was not our organizing model, and friendship was not as necessary or as beneficial as mutual respect, compassion,
patience, openness, and consciously structured group process.[4]

WYMSM never produced the software we hoped would be the primary
result of our collaboration. In interviews after the disbanding of the group,
several members expressed frustration with our inability to complete major
projects, like the Beat the System software game. For example, when I asked
Cosandra Jennings how WYMSM could have been better, she replied, "I
think FOCUSING—focusing on what we did last, crossing things off a list,

Figure 7.2
Research on technology and poverty. Front to back: Kelly Klein, Jenn Rose, Ruth Delgado Guzman (with her back to the camera).
Photo: Pat Dinkelaker

and then knowing what we had to do next. It did make me discouraged—I felt like we were going in circles. I miss the group a lot, but it was so hard because people had so many other things that they had to do." Similarly, Zianaveva Raitano commented, "We certainly need[ed] structure, especially with scheduling. Making sure that everybody knows that their role is so important—everybody needed to see that and have the devotion to want to make it through. Because we were making progress, we were meeting goals. Sometimes we would go off and start talking about [other] really important issues, but we were on the right track."

WYMSM members pointed out that being consistent and maintaining focus are particular challenges to organizing as poor and working-class women: when time is tight and you have to make a choice between meeting your basic needs and attending a meeting, it is difficult to maintain a commitment to political organizing, no matter how much you want to bring about change. Jenn Rose mentioned, for example, "We needed a lot more time than we had. If everybody had more time, to actually sit and work on the project, we could have gotten a lot more done. Poor people don't have time, because they have to work so hard just to survive. . . . I don't know how we could make that better. Certainly a living wage would help." This is a pressing double-bind, indeed: women in the YWCA community need time and resources in order to organize for the time and resources they need.

One way WYMSM met this challenge was by offering participants a stipend of between $6.66 and $7.50 an hour for participation in the group. Perhaps the most unequivocal lesson WYMSM offers is that cross-class, inclusive participatory research cannot take place without some form of direct remuneration. As Sanford Schram argues, social change relies on careful attention to the day-to-day politics of survival.[5] The political and material work of providing for mutual survival is often underestimated or outright ignored by political theorists and activists, perhaps because it is less sexy than the confrontational politics of street protest, and its outcomes are less clear than those of organizing that directly engages with formal politics through lobbying, voter registration, or letter-writing campaigns. Perhaps, too, this important form of political engagement is ignored because it is work largely done by women. But for political activity to contribute to broad-based equity, it must "meet people where they're at," in terms of their immediate circumstances and concrete needs. As Schram argues, the politics of survival is necessary not only because material supports are indispensable to continued political organizing, but also because active engagement with others, embedded in the realities of their everyday lives, is a crucial source of knowledge that can contribute to democratic justice (Schram 2002, 37).

The stipends had powerful impacts on the affective bonds of the group, and on members' commitment to the process. Rather than compromising their objectivity or attracting members with base material motives, the group's participatory practice and the symbolic message sent by the stipends, more than their monetary worth, increased trust and motivated WYMSM members' continued commitment to social change. Zianaveva Raitano explained,

Virginia: I remember you saying to me once that WYMSM really helped you feel like your opinions and your beliefs were respected.
Zianaveva: I do. I'm so grateful for that experience. You made everyone feel like they had the time and the space to do that, that their opinions were valid.
Virginia: And you said another interesting thing: that one of the ways your knowledge was being validated was because you were being paid for your time.
Zianaveva: Yeah! That was really important. Because it goes beyond just volunteering your voice. It's somebody acknowledging that there is a financial need, aside

Box 7.3
WYMSM Member Profile, Zianaveva Raitano

"I learned by being in WYMSM that I can do anything."

Zianaveva Raitano was a creative, generous and hard-working member of the WYMSM team. Though she characterized herself as being "at the beginning stages" in her political views, she was remarkably active politically, and always advocated the radical route—protest in the streets—over reform or politics-as-usual. She was an antiwar activist and involved in environmental, racial justice, and community health issues. This level of political involvement is even more extraordinary considering that Zianaveva worked several jobs in health care, data entry, and social services—often several at once—for the entire time I knew her.

I met Zianaveva, a young, working-class African American woman who grew up in public housing, when she played the banker in the Life in the State of Poverty simulation. One of my favorite pictures of the event is Zianaveva sitting behind a folding table and counting a massive pile of play money, her face lit up by a beaming smile. Shortly after that event, she attended the final sessions of the Women's Economic Empowerment Series, and then joined WYMSM. Her good humor animated our meetings, and her sharp intellect contributed enormously to this project.

In our interview in 2003, she told me that her favorite part of WYMSM was the feeling that "There was no limit to what we could do as a group, if we put in the work. In my life, I've always felt restricted and limited in the things I thought I could do. But I learned by being in WYMSM that I can do anything."

Zianaveva was the first to point out that YW residents were increasingly unable to find independent housing in Troy because the economic changes of the early 2000s were forcing poor and working-class people out of the area. She argued, "Rent control is a big thing. These private owners are hiking up their rents. The working class and the poor are going to suffer because they can't afford it. We need to give landlords an incentive to not jump on the bandwagon and hike up their rents." Ironically, after WYMSM disbanded, Zianaveva was one of the working-class people who could no longer afford to live in Troy. So when she left the YWCA, she left the area, moving further downstate.

Nevertheless, we stayed in contact for many years. We talked on the phone occasionally, and I wrote her a few letters of recommendation, as she had worked for me on the Women's Resource Directory project. I still have a message from her on my answering machine. Unfortunately, the last time she called was an incredibly busy time of my life, and in the several weeks it

Box 7.3
(continued)

> took me to return her call, her phone was disconnected. Despite putting an advertisement in the regional newspaper and writing to her family home, I was unable to find her to complete a first-person profile before this book went to press.
>
> Another irony? Zianaveva was the one who suggested using member profiles in the book. Yet she is the only surviving member of WYMSM I was unable to find.
>
> *Drawn from an interview on November 4, 2003, and field notes, 2001–2004.*

from just *talking* about poverty and people struggling. I mean, here's a group that actually has something to offer financially. That was instrumental in sending the message that our time was being valued. And that sends a powerful message, as opposed to just coming in and volunteering and donating your free time. You felt validated, you felt important.

Virginia: It wasn't a ton of money! Did it play a more important role as a symbol than it actually did in supporting you and making your time available?

Zianaveva: That's exactly what I'm saying. When we're talking about poor people, and social justice, here's a financial benefit. You're not just talking about it. You're *doing* something. [I]t wasn't just talking, politicking. We actually had this part of the group that spoke loudly about economics. It wasn't a lot, but it was a whole lot to me. . . . That's an example of some kind of justice. I've been in groups before, and nobody ever put forward the idea of offering something to people for their time. This is the first time I've ever seen that happen. It was revolutionary!

When engaging in participatory research, some scholars choose not to pay research collaborators out of concern about impacts on research validity or conflicts that might arise in community-university and interpersonal relationships. For example, a fellow scholar once explained to me that she does not give stipends to research participants for fear it would "set up inequalities among individuals in the community." Similarly, when writing a grant application for a WYMSM research project, I received a review that expressed concern that paying participants would compromise their objectivity.

When I addressed these concerns directly with other WYMSM members, they argued that payment has the opposite effect, strengthening the objectivity of our research and educational efforts by maximizing the diversity of the pool of possible participants and by proving the university's commitment to—and accountability for—community processes. WYMSM members questioned instead the ethical and empirical validity of work that *fails* to provide for broad-based participation and exchange by meeting the very real material needs of people at the "street level." During our interview, Jenn Rose spoke at length about the difficulties women in the YWCA community faced attending meetings, and she and I reflected on the role stipends played in our process:

Virginia: Was paying people for their time something that made [this project] a little more possible?
Jenn: Yes! If you want poor people in your group, they have certain needs. Get a clue. Get out of your high towers! Come down to street level. Maybe if [academics] did more of that, maybe things would get fixed.
Virginia: So do you think the money biased you in any way?
Jenn: No! No. The money showed us that you actually valued the person's time. And not just that it was money. But it wasn't just like . . . "I'm doing this, please come in and talk. Thank you. Goodbye." Not like "I'm getting something out of this, but you're not going to get anything out of it."

Material support for, and validation of, members' participation was absolutely central to the success of the project, and WYMSM was committed to providing remuneration for women's often unacknowledged community-building labor. Nevertheless, managing the stipends was often difficult, emotionally loaded, and interpersonally tense. During WYMSM, I was a research assistant who managed organizational details of the project, including responsibility for managing and distributing money. This was a task that my own unacknowledged discomfort with my middle-class background made intensely complicated. Because I had internal conflicts about the money and my class position in the group, I did not always communicate effectively and clearly about WYMSM financial status or our procedures for payment. The lack of transparency made financial transactions seem arbitrary, and led to interactions—people asking for advances, confusion over process and amount—that damaged the group dynamic. Though I cannot verify it, because she passed away before the project was complete, I believe that this tension was one of the reasons that "Coffee," one of our earliest members, chose to leave the group.

The stipends were also complex from a structural point of view. Many women in the YWCA were receiving some kind public assistance, and paying them by check would result in a 100 percent penalty from social services—any dollar they earned would be taken out of their benefits. On the other hand, it was impossible to meet the reporting requirements of a federal granting agency and a university by giving cash stipends; the university had to issue checks. Our solution was often to give workshop participants and WYMSM members gift cards as payment, which gave rise to its own set of contradictions. For example, we held a workshop on the global assembly line, discussing the role of big box chain stores in breaking unions and driving down wages. We then paid participants with Wal-Mart cards, as it is the only retailer they can easily reach on the bus! Work with WYMSM was often a powerful lesson in living with contradiction, with "good enough" solutions, and with resisting ideological Puritanism. In order to meld the politics of survival and the politics of social change, you often need a sense of humor.

Catalytic Community Building

WYMSM undertook and achieved a number of important projects, but our original purpose was to complete a software game, to produce a technological product. How, then, do we judge the success of the group? The framework that I have developed thus far requires seeing popular technology as a way of engaging in the citizenship dilemmas of the information age as well as a way of developing empowering "technologies for people." Thus, while the completion of technology design projects is important, popular technology must be judged on its success in helping participants recognize, reframe, and engage their political identities and power.

The atmosphere of equal exchange and mutual education created by our participatory process and the material and symbolic support provided by the stipends, WYMSM members explained, is the soil from which social movements grow. Though we did not accomplish all of our original goals, WYMSM members felt that we acted as a catalyst for personal, organizational, and political change. Many WYMSM members talked about the role of the group in catalyzing other social justice activities and described their hopes that others would pick up where we left off. Cosandra explained, "Hopefully [new people] will get farther along than we did. . . . It's like we put that platform down, [planning] how we were going to make this game. We had great ideas. But we [only got] so far." Jenn Rose insisted that "we brought people in. . . . [WYMSM] was like a catalyst." WYMSM's work did, in fact, catalyze and support several other community projects.

A welfare rights and economic justice organization called Our Knowledge, Our Power: Surviving Welfare (OKOP) grew out of WYMSM. We had our first meeting at the YWCA in July 2005, received a grant, and moved to a new meeting space in nearby Albany. We continue to operate as an independent grassroots group with more than fifty members in the Capital Region, including some of the original members of WYMSM. The Hudson-

Figure 7.3
Women's Economic Empowerment Series session on women's work. Visible, left to right: Nancy D. Campbell, Christine Nealon, Julia Soto Lebentritt, Jenrose Fitzgerald, Jenn Rose.
Photo: Pat Dinkelaker

Mohawk IndyMedia Center moved into the YWCA shortly after several of its members participated in the ROWEL simulation with WYMSM, produced an award-winning video, and used the proceeds to buy its own space in North Troy, now an important hub of local cultural and political activity known as the Sanctuary for Independent Media. Rensselaer faculty members and students, including Nancy D. Campbell, Tamar Gordon, and Branda Miller, continued to undertake projects with YWCA residents through the community technology lab, including a popular technology workshop that culminated in "Through Harriet's Eyes," a group presentation on Harriet Tubman and local history in the area's well-regarded Underground Railroad Conference in 2003.

Our success in inspiring other projects at the YWCA, and in the broader community, was an effect of our process focus. When I asked her how we should judge the group's efficacy, WYMSM member Julia Soto Lebentritt commented, "That's the thing to look at—the process. That's community. That's community building. And empowerment." When I asked Cosandra

Jennings what the perfect future of WYMSM looked like, she responded, "Bigger. More out there in the community. More willing to make social movements. That's what our whole purpose is."

Engaged Objectivity

WYMSM's participatory approach created a context of trust, reciprocity, accountability, and transparency in which members' feelings of entitlement to political articulation grew and strong bonds within the group were forged. Our participatory process also created rich, robust, timely knowledge: both practical knowledge that helped us plan political action in a specific context and more abstract, generalized theories about the relationship between technology and politics, the information age and inequality. The unique insights represented in this book—ideas that sometimes run counter to accepted social scientific and public policy understandings of technology and inequality—are a direct result of the collaboration and engagement at the heart of the process.

Throughout the book, I have tried to represent the data WYMSM and I gathered in its context and as it arose, in dialog, with no one voice—even mine—establishing the final, authoritative version of the facts. This may strike some readers as non-neutral or insufficiently objective. But I think we can expect more from social science than transparent description of social reality. As Bent Flyvbjerg has provocatively argued, social science has not yet demonstrated that it can predict what human agents or societies will do, given a certain set of conditions. Nor should it try to achieve the predictive aims of the natural sciences. Knowledge of the social world is always context-dependent, historically contingent, and partial. Instead, Flyvbjerg argues, social science should jettison purely instrumental goals to specifically focus on clarifying values, interests, and power relations (Flyvbjerg 2001). Participatory approaches to research are particularly well-suited for collaboratively answering the kinds of questions Flyvbjerg asks: Where are we going? Who gains and who loses, and by which mechanisms of power? Is this development desirable? What, if anything, should we do about it?

Robust knowledge is created when investigators refuse to produce unlocatable, irresponsible knowledge claims under the guise of neutrality.[6] Participatory approaches to research produce what Sandra Harding calls *strong objectivity*[7] by multiplying the diversity of inputs in a research program and situating knowledge claims in their context. Participatory approaches to research produce knowledge that is not neutral but is highly objective, rigorous, and generalizable. Such approaches integrate the points of view of multiple analysts, provide ways of triangulating between the standpoints of participants from a wide variety of social locations, and increase accountability and transparency in research. I like to think of participatory action research as developing "engaged objectivity": insights are strengthened each time they pass through the mind of another analyst; interpretations are validated each time the results of research are tested in action in the real world.

Engaged objectivity thrives in polyphony, not in the sterile bell jar inhabited by the neutral academic observer. Genuine objectivity can only be achieved through a plurality of partial and contingent perspectives. As Donna Haraway argues, objectivity is not about disengagement but rather about "how to have *simultaneously* an account of radical historical contingency for all knowledge claims and knowing subjects, a critical practice for recognizing our own 'semiotic technologies' for making meanings, and a no-nonsense commitment to faithful accounts of a 'real' world" (1991, 187).[8]

Transforming social relations is work that we must do together. Transformation occurs when we all bring our full selves to the work, firmly rooted in and understanding the powers, privileges, and limitations of our particular social locations, striving to be accountable for our thinking, our actions, and the democratic process in all its complexity. The co-production of knowledge is enormously complicated—but it is possible. As María Lugones and Joshua Price argue, only by recognizing the complexity of communication can we create knowledge that "does not reduce us to being simply images of one another" (Lugones and Price 1995, 124). Only complicated knowledge and difficult conversations can give rise to social justice. And so, participation is *hard*. Participation requires that we co-create knowledge while recognizing that everyone's perspective is incomplete, that power relationships systematically distort our communication, and that there are limits to understanding the lives of others.

Critical Citizenship and Cognitive Justice

Delivery-oriented, distributional solutions to the equity problems of the information age fail to interrupt systems of epistemological privilege and power, and thus, current high-tech equity policy tends to displace women in the YWCA community, discipline and survey their decision making, and obscure their lived experiences. In this context, conventional liberal humanist ideas about community involvement, participation, and "local knowledge" are not enough to tip the information age toward social justice. From WYMSM's point of view, even the emphasis on new technologies providing marginalized people a chance to come to "voice" seems naive.[9] Community involvement often devolves to mere consultation, and lacks accountability and transparency when it comes to actually making decisions. The rhetoric of participation can be cynically deployed to enroll oppressed people into the process of their own domination, burdening them with extra responsibilities while failing to shift power relations or patterns of material inequity. "Local knowledge" can be fetishized, resulting in essentialist narratives about difference that minimize opportunities for alliance building and political mobilization. "Voice" is too often forced into political scripts that demand that poor and working-class women speak only as public witnesses to their deficiency or debasement, and not as active agents in their own complex lives. The challenge of popular technology, thus, is an epistemic challenge. WYMSM's call is a call for cognitive justice and the liberation of knowledge.

Cognitive justice and epistemic liberation must start by acknowledging that all social locations provide *both* uniquely clear vision (in some areas) and specific barriers to insight (in others). While none of us are perfect interpreters of our own experience, the gaps in the knowledge of the powerful are systemic and significant, as my own example in this book makes clear time and time again. Socially or culturally sanctioned ignorance that supports the status quo is created and maintained by institutions such as schools, workplaces, families, and neighborhoods. The standpoints of the marginalized, the knowledge that is generated from their unique and particular social locations, thus offer important epistemological resources in any quest to create a more just world. As Sandra Harding has written, "Marginalized groups have interests in asking questions [about power], and dominant groups have interests in not hearing them" (Harding 1998, 151). This is the case both in knowledge arenas considered neutral, such as science and technology, and in knowledge arenas that are obviously embedded in power relations and entrenched interests, such as knowledge dealing with poverty or other forms of systemic social violence.

We must be cautious when making claims for epistemological privilege, however, as they can quickly become arguments for epistemological superiority—reinscribing a hierarchy of knowledge that is profoundly antidemocratic. Knowledge that arises from a particular social location—what feminist scholars call a "standpoint"—is an incredibly useful resource when it is seen as akin to DuBois's "double consciousness." For marginalized people, understanding the world is characterized by multiple points

Box 7.4

WYMSM Member Profile, Nancy D. Campbell

"Wouldn't we be smarter as a society if we had more of this co-production of knowledge?"

At the time we started WYMSM, I was a recently divorced mother of one, going through all these transitions. It was an unstable moment. I had been teaching in Women's Studies, and then I had switched to a place that felt like a hostile climate for women, and I felt destabilized by that. I was looking for a community of women to keep my feminist sensibility together. I was a "subversive" academic who was interested in whether it was possible to remain deeply involved in social movement work and also meet my overwhelming responsibilities as a teacher and as a writer.

I was going through a lot of change in my life at that point. At the time, I was trying to make a transition from being somebody who was really defended and shielded, in part because I was masking great insecurity. I always had this sense that I'm not really meant to be in the academy. I step onto campus to this day and I think, "People like me are not really meant to be here." At that point in my life, I didn't feel like I could admit that. I felt like I had to appear invulnerable.

WYMSM helped. It was kind of like jury duty, where your power and privilege by virtue of being an academic doesn't really work for you and can actually work against you. The nice thing about WYMSM was you got to be who you were. You didn't check your differences at the door; there was no discourse of neutrality. It's not that other members of WYMSM didn't *know* that I had a Ph.D., it's more that it was *immaterial*. In another group I work with on social justice issues, we have very different styles and languages, and we interact a great deal across class and across race. But there's a discomfort and a formality, something we didn't have in WYMSM. It wasn't terribly formal; it didn't feel weird.

I remember ROWEL really favorably, because it generated such a sense of excitement in participants. Having the children from Taylor School there, along with WYMSM, and HANNYS, brought together a lot of constituencies, and it felt like there was momentum. It was a big splash moment. What I remember most about the workshops themselves—both the WYMSM meetings and the Women's Economic Empowerment Series—is the sense that we were making space for creativity in everyday life. It gave everyone who participated a sense that they were connected and doing valuable work—a way to express themselves. It was liberating.

The academy might discount work like that, but the connections felt very real. We were building methodologies for doing things; there are continuities between what we all do now and what we were doing then. Those connec-

Box 7.4
(continued)

tions are not so much broken now as they are transformed. It was important and empowering.

Five years from now, I still want to be in the academy, but I want to be spending more time with my kids, be a little more relaxed, and only be doing things I want to do. I study how drugs and drug policy affect American life. Many people tell me I should get away from studying drugs, but they are compelling to me, because drugs are technologies of the self. [Thinking about drugs] is a place where people most acknowledge that there is a big gap between what's supposed to be and what is. [Drug users] fall into that gap. We live in a world that is crazily transforming: increasingly insecure, undermining of human potentials of all kinds, and this is a way of making a sanctuary. For some reason drugs allow you a certain kind of insight into that or a way to cope with it. I feel like it's a very fruitful place to talk to people. They share absolutely amazing things with me when they find out what I do. It's a way of very quickly getting into very real conversations. I feel very privileged to be in that kind of relationship with people.

I want readers to know that sometimes academics are regular people. Also, that there is something meaningful about collective expression that is different from what I put together by myself. I'm not sure before the WYMSM experience that I really got that. Wouldn't we be smarter as a society if we had more of this co-production of knowledge?

Based on a conversation over coffee, April 18, 2009.

of view: those that arise from your counterhegemonic subject position as well as those that arise from the subject position you must inhabit in order to exist successfully in dominant society. Standpoint is less useful when it is seen as essential knowledge that adheres to a particular kind of person—"women's knowledge," for example. Standpoints are relational and learned, fought for and won. They are not demographic, biological, or otherwise natural positions we inhabit (Wylie 2003). All social locations are partial, non-innocent and complicated—the danger lies in losing the specificity of socially located knowledge by collapsing complex relationships onto simple demographic categories of identity.

Participatory, collaborative approaches allow us to focus on the social structure that creates subject positions rather than on the positions themselves. Popular technology projects call for engaged, collective inquiry across difference, because *all knowledge is local knowledge* and every stand-point can produce important information for analyzing our shared world.[10] For WYMSM member Julia Soto Lebentritt, what worked best about WYMSM was the cooperative process we developed to create collective knowledge across difference. It is in the collaboration of a diverse group of participants that she feels true individual *and* social empowerment lie. She remarked in our interview, "It was beautiful that we were from differ-ent walks. We had different ages, different experience. But we were all on the same level. It was really across different cultures, diversity . . . interact-ing positively . . . almost like a family gathering of . . . what we were pas-sionate about, and what we felt should be done." In short, there is a lot of work to do, and we all have a role to play in creating a more just world.

According to Shiv Visvanathan, cognitive justice demands both cogni-tive representation and broad-based participation in knowledge-making practices. He reminds us that epistemic plurality is central to democracy because, in any system, choice cannot exist without alternatives. Democracy *within knowledge* is as crucial as democratic process in institutions or deci-sion making. Thus, "cognitive justice [is] the constitutional right of differ-ent systems of knowledge to exist as part of dialogue and debate. Cognitive justice recognizes the plurality of knowledge systems [and also] the relation between knowledge and livelihood or lifestyle. Cognitive justice goes beyond voice or resistance to recognizing the body of knowledge within which an individual is embedded" (Visvanathan 2005, 92–93). Cognitive justice demands that competing understandings, analyses, or frameworks be allowed to coexist within the knowledge ecologies that support them. This means neither "museumising" marginal knowledges by freezing them in time and place as static and essentialist nor ignoring entrenched power systems that create very real knowledge hierarchies by identifying whose knowledge is "expert" and whose is "lay," whose experience matters and whose is marginalized or obscured.

The incredible richness of the YWCA community offered important resources for creating cognitive justice. Far from typifying some kind of monolithic culture of poverty, the YWCA community offered vibrancy, diversity, and resourcefulness. YW residents are seventeen to seventy, have a fifth-grade education and multiple master's degrees, live at the YW for six weeks or twenty-six years. They work as nurses, data specialists, cooks, counselors, and artists, as well as being mothers, lovers, wives, sisters, and friends. The walls of the YWCA are highly permeable—in the years I spent in the community, I interacted with residents, staff, administrators, local artists, activists, community organizers, day care parents, board members, and others who use the YWCA facilities, participate in its activities, and enrich its lively culture.

Women in the YWCA community have absorbed more than their fair share of the stresses and strains produced by the collision of hypercapitalism and neoliberalism. They are "those hardest hit by privatization of services in which they were previously employed, those who are in the front line of negotiating for welfare services, and those who voluntarily give their unpaid labor for the collective good" (Ledwith 2001, 172). Because of this position and their experiences with the extractive and exploitative faces of IT, women in the YWCA community were able to develop incredibly insightful analyses of the relationship among technology, politics, and poverty in the United States.

The research we did together did not simply gather local knowledge. The experiences of women in the YWCA community, many of whom work, play, and live within a tightly circumscribed five-mile radius of the building, nevertheless illustrate how the local is enmeshed in transnational processes, and vice versa. As my collaborators pointed out to me, poor and working-class women across the globe serve as the canaries in the coal mine of the information society, test subjects for social control in an era of government devolution and neoliberalism. They also provide much of the unpaid and underpaid labor—in the service industry, as caregivers, as data processors—on which the megalith of the information economy rests. The experiences of a small group of women struggling to meet their basic needs in a small city in the American Northeast can tell us a lot about the broader relationship between technology and social justice.

Cognitive justice in the information age demands that multiple knowledges be recognized as arising from specific social locations and be integrated into decision making. What is needed for better, more just decision making about technology, as Visvanathan argues, is a *parliament* of knowledges. This parliament will not build itself. Building open and inclusive political institutions is *our* job, all of us. What is needed for high-tech equity is not better public understanding of science and technology or a transfer of technology and knowledge from the haves to the have-nots. Women in the YWCA community understand the impacts of science and technology on their lives perfectly well. What is needed for high-tech equity is an agenda for understanding and transforming the real world of information technology that we all share.

Conclusion: A High-Tech Equity Agenda

The information age behaves less like Noah's flood, washing away the evils of industrial capitalism and leaving behind a playing field that is clean, smooth, and level, and more like Hurricane Katrina. Katrina revealed, with great violence and human suffering, the desperate inequalities that underlie American society, inequalities sedimented over decades through bad policy, human indifference, and oppressive institutions. That the blinders of the privileged were only temporarily torn away during the Gulf Coast catastrophe should remind us that the work of social justice must be conscious, daily, personal and collective work. It is ongoing, terrifying, glorious, immense. It takes clear vision. We know the flood is coming. We know the levees might break. The rising tide of the information economy does not lift all boats: it sinks some, destroys others, and drowns the boatless.

In September 2009, President Barack H. Obama came to Troy to deliver a speech about retooling the American economy for the high-tech future by nurturing innovation, supporting basic research, and increasing college attendance, especially in institutions like Hudson Valley Community College, where he spoke. Acknowledging that regions like upstate New York "have been dealing with what amounts to a permanent recession for years" (Obama 2009, 2), he laid out a strategy to foster new jobs, new businesses, and new industry. In the speech, he reaffirmed his administration's commitment to investing $100 billion in high-tech classrooms, health information technology, and sustainable energy. He laid out plans for an American Graduation Initiative, which would increase Pell Grants and tax credits for education, reform the student loan system, create a GI bill of rights, and strengthen community colleges. He promised to bolster entrepreneurship by increasing broadband capability across the nation, keeping the Internet open and free, and investing in basic research in the private and public sectors.

I am deeply sympathetic to many of his goals. But, with the exception of a brief comment that true economic recovery must include "sustained growth and widely shared prosperity," President Obama's energizing talk said very little about struggles for justice in the information age. Though public policy during the last three presidencies offered a variety of different solutions to high-tech inequity, each administration insisted that the problem was the same: poor and working-class people lack access to high-tech products, education, and jobs. This assumption is incorrect, and leads to misguided analysis, policy, and activism.

A just information age is not beyond our reach. But we must seek innovation *and* equity, economic growth *and* economic justice. Much of the current high-tech equity policy and scholarship dismisses the resources of poor and working people, either mourning them as inevitable victims of progress or seeking to retool them to "fit" into the new economy. Neither approach could have unleashed the powerful analysis or considerable resources of women in the YWCA community. Neither will result in an information economy that respects and protects human rights, lessens inequality, and invigorates democratic governance. We must create alternatives. Popular technology can be one resource for defining a high-tech equity agenda that minimizes risks, more fairly shares rewards, and ignites the passion and energy of all people in robust democratic processes.

Our existing models of technology education and policymaking separate technology from its context and discount the importance of human experience, agency, and decision making. Existing models of technology training, primarily vocational, render structural forces invisible and constrain our thinking about the future. Vocational training can be important and useful, but it simply cannot produce the kind of critical consciousness we all need in order to imagine alternatives to the status quo. Popular technology examines science and technology as deeply embedded in our everyday lives, imbricated in systems of power and privilege and responsive to human values and efforts. Popular technology offers all of us the tools and opportunity to become more critical in how we think about the relationship among technology, society, politics, and inequality.

While it is difficult to programmatize popular technology, as the central goal is to be responsive to specific struggles in their social context, it is possible to suggest ways of evaluating technology projects in terms of their ability to foster critical technological citizenship. There are some basic questions you can ask about your next technological project, action, program, or policy to align it with the goals of creating a technology "for the people, by the people." I hope the handout in box 8.1 will help you evaluate your own undertakings.

Box 8.1

Is It Popular Technology?

Popular technology reminds us that technology is not a destiny but a site of struggle. Inspired by the tenets of popular education, the approach is grounded in the belief that people closest to problems have the best information about them and are the most invested in developing smart solutions. Popular technology sees *all people* as experts in their own experience of IT and the high-tech economy, and liberates their knowledge, analysis, and activity to create a more just and sustainable technological present for everyone.

Does your technological project, action, program, or policy:

Resist Oppression?

Exploitation   Does your undertaking acknowledge and resist the transfer of the benefits of the labor of oppressed groups to other social groups?

Marginalization   Does your undertaking include the experiences, efforts, and input of people who are excluded from full participation in social life?

Powerlessness   Does your undertaking explicitly aim to expand the authority and status of oppressed and marginalized groups politically, culturally, economically, and socially?

Cultural imperialism   Does your undertaking resist the impulse to universalize the culture, values, and experiences of one social group as normal, natural, and correct?

Violence   Does your undertaking actively discourage violence as a systemic social practice, whether it is physical, structural or economic?

Draw on Difference as a Resource?

Cognitively   Does your undertaking begin from the assumption that people closest to problems know the most about them? Does it include explicit structures and procedures for drawing on a variety of different perspectives, from different social locations?

Culturally   Does your project recognize that the collective experiences, beliefs, and values of different groups of people are immensely valuable in analyzing social issues, making good decisions, and building strong organizations? Are you building structures in your project to get beyond "diversity" to learning and acting for equity and justice for all people?

Institutionally   Is your organization explicitly committed to protecting human rights and fighting for sexual, racial, gender, and class equity? Do you have explicit goals, policies, and accountability mechanisms in place for ensuring diversity in your project, organization, or institution?

Engage in Participatory Decision Making?

In agenda setting   Who identified the issue that the project is taking on? Might there be ways to broaden the agenda—to establish how collaborators understand and define the issues that most affect their lives?

Box 8.1
(continued)

> *In design*   Does your project see design as a dynamic process of defining terms and clarifying principles, as well as setting plans for concrete action? Does it acknowledge, and have processes for dealing with, the conflicting goals and values that the design process inevitably uncovers?
>
> *In implementation*   The impact of your project will largely be determined by those who implement it. Have you included them as full partners in all stages of the process? Is there a place for their input to feed back into an iterative design process?
>
> *In evaluation*   Action that is not followed by reflection is only half done. Unfortunately, evaluation is often slighted when funding, time, or energy runs out. Do you have an evaluation plan? Have you left enough time, energy, and space to do vigorous evaluation at the end of each stage of your program?

Perhaps the most important lesson I learned from popular technology is that we must judge the success and efficacy of our participatory projects by their ability to foster growth, in ourselves and in others who work alongside us, rather than their achievement of narrowly defined deliverables. Programs and projects that start from a participatory standpoint often take unexpected turns, so flexibility and responsiveness are key, as is an ability to take a broad and critical view of one's own process. One of the most surprising outcomes of my work with WYMSM, for example, was how quickly our agenda broadened to encompass justice issues outside the narrow confines of what I originally saw as technological. By broadening our focus beyond developing technological artifacts and skills and understanding technology and the information economy in the context of participants' everyday lives, WYMSM opened a way to think more broadly about what social justice means in the information age.

When I asked women in the YWCA community what advice they would give the mayor of Troy about building a more just Tech Valley, their answers were not what policymakers might expect. Zianaveva Raitano advised the mayor to seek better information about what was *actually* happening, saying, "Tech Valley is supposed to be the next big thing, so the press should be reporting regularly on what's actually going on: hiring rates and unemployment rates!" Women in the YWCA community critiqued then governor George Pataki's "genius plan" to bring high tech upstate, arguing that it was driving up rents and driving out working families. They explained that displacement, impoverishment, unemployment, and environmental destruction are not worth a handful of high-tech jobs benefiting highly educated people. Cuemi Gibson's response was typical:

Virginia: If you had the ear of the mayor of Troy, what advice would you give him for developing a socially just Tech Valley?

Cuemi: Put plants on the bus line! Provide education and training; provide fair employment. Affirmative action! Have some Black supervisors, female supervisors. Work with an open-door policy. Be fair in what hours people work. Have a day care center. Try to have stress-free environment. Provide employee assistance, like rehab and counseling for long-term employees. [Z]ero tolerance for racism and sexism. [F]air employment. I'd like a fair pay rate. Not a minimum wage, a *living* wage.

Hearing women in the YWCA community's answers, I would try and refocus. "OK," I'd say, "so definitely more buses, less racism, fairer employment. But what about technology and high-tech development *specifically*?" And they'd answer, repeating themselves, looking at me as if I had lost my mind, "More buses. Less racism. Fairer employment." Technology is about people's whole lives; it is the house we all live in. It is deeply enmeshed in politics, economics, citizenship, and day-to-day survival. Women in the YWCA community saw technology as so deeply woven into existing structures, systems, and institutions that it could not possibly be extricated from other agendas for social justice and reform. They refused to see IT as separate from its context. They lived in the real world of information technology every day, and their advice grew out of their specific, concrete knowledge and needs. In their spirit, I offer the following agenda for creating equity in the information age.

A High-Tech Equity Agenda

1. Protect Workers in the Lower Tier of the High-Tech Economy

The decline in the manufacturing sector and increase in information and service sectors are not bad news because jobs in manufacturing are, in and of themselves, better jobs. The problem is that the rapidly decreasing manufacturing sector has traditionally offered unionized work, relatively stable employment, and a living wage. Work in the information economy—in both the high-tech and the service sectors—tends to be less reliable and lacks institutions for effectively bargaining on workers' behalf, especially unions. Union members in the United States make 30 percent more than their non-union counterparts and have greater access to health care benefits, short-term disability coverage, and life insurance (Jobs with Justice 2009). The current economic crisis makes it all too obvious that many workers have lost their power to bargain for fair wages, workplace safety, robust pensions, and job security.

Though consumer service, caretaking, and other bottom-tier jobs in the information economy have often been characterized as difficult to unionize, amazing campaigns by and on behalf of these workers have arisen in recent years. For example, the Service Employees International Union (SEIU) is the largest and fastest-growing union in North America, boasting more than two million members who provide for America's health care, public services, and property services needs. SEIU is also the country's most diverse union: 56 percent of SEIU members are women, 40 percent are people of color, and SEIU represents more immigrant workers than any other union in the United States (SEIU 2009).

Domestic workers have been organizing for their rights in increasing numbers in the past decade, as well. Exempt from most labor protections, hidden away in private homes, and often facing language or immigration barriers, domestic workers are both absolutely central to the information economy and extremely vulnerable to workplace abuse and mistreatment. Domestic workers are explicitly excluded from the National Labor Relations Act and thus are unable to form labor unions. Nevertheless, Domestic Workers United, formed in 2000, has organized hundreds of Caribbean, Latina, African, and African American nannies, housekeepers, and eldercare workers in New York. DWU provides a stellar example of workers organizing to protect their rights and assert their dignity, undertaking research in their communities, fighting for justice for exploited workers, and exerting political power. Their proposed domestic worker's bill of rights seeks to provide basic workplace protections that domestic workers currently lack: overtime pay, days off, advance notice of termination, severance pay, cost of living increases, health care, paid sick days and vacations (DWU 2009).

2. Take High-Tech Industries off Welfare

Billions of dollars of federal, state, and local money are spent yearly to attract high-tech industry to struggling areas like the Capital Region. But it is unclear what impact these subsidies have on companies' decisions to relocate, and there is even less evidence that these subsidies have a net positive effect on local and regional economies. In the Capital Region, for example, more than $10.5 billion in public and private money has been promised since 2000 to subsidize a few key high-tech players, including IBM, General Electric, and Advanced Micro Devices (AMD). Much of this money has been allocated for developing or supporting research institutes—Albany NanoTech and General Electric's Global Research, for example—or for building or updating chip fabrication plants.

New York State has offered AMD/GlobalFoundries a $1.2 billion package of cash incentives and tax breaks to locate a new chip fabrication plant in Malta, New York. In addition, massive community and public investments in infrastructure have been provided to build Luther Forest Technology Park, where the $800 million plant will be located. After two years of intensive construction work, which will create 1,500 temporary jobs, this plant will employ just 1,400 people, at a taxpayer expense of $857,000 per job. It is quite clear that these massive subsidies are the only thing attracting AMD/GlobalFoundries to New York State. Most chip fabrication plants are built in areas with weak environmental controls and worker protections—one reason why Austin, Texas, was such an attractive option for high-tech manufacturing industries. New York State, with its traditionally strong labor and environmental regulations, has had to fight hard and spend big to bring these industries to the region by promising mega-incentives, some of the largest public investments in private industry ever recorded in New York.

Many subsidies are justified in hopes that they will draw new businesses to the area and create new jobs, as in the case of AMD. But most subsidies to high-tech in upstate New York go to General Electric, which is headquartered in Schenectady, and IBM, which has operated in East Fishkill since 1962. Since 2000, $1.365 *billion* in public money has gone to keep IBM at the East Fishkill site and to encourage the company to expand, for example, by creating a research and development center in Albany and a chip packaging plant near Utica. Since 2000, mega-incentive investments in Tech Valley have netted the area 9,400 jobs, at a cost to taxpayers of roughly $135,000 per job. This does not include the $1.2 billion for AMD's promised 1,400 jobs; if we include the AMD subsidy, the cost per job for 2000–2010 would go up to $225,000.

The incentives and tax breaks provided to the high-tech industry in New York, averaging $1.4 billion a year since 2000, would more than pay the state's contribution to public assistance and social services, including food stamps, cash assistance, child care subsidies, welfare-to-work programming, summer youth employment programs, and domestic violence prevention services. We need to do a true cost-benefit analysis of this regional development strategy. Ill-considered business subsidies create perverse incentives. High-tech research and development subsidies sometimes make sense, but before handing over billions in taxpayer money, we should ask a few basic questions: Can the company afford to pay for its own research, development, and expansion? Has the company looked for sources of private funding before attempting to access taxpayer money? Does the price we pay per job make sense? What other benefits are these investments likely to bring to the area? What other burdens do these investments place on our communities?

3. Respect and Reward the Work of Care

As Nancy Folbre eloquently argues, "the invisible hand of the free market depends upon an invisible heart of care" (2001: vii). The information economy drives increases in employment in the human and consumer services industries, amplifies the vulnerability of many American families, and exposes us all to more of the shocks and strains of volatile continuity. Women have traditionally borne disproportionate responsibility for the invisible heart by performing family and community caretaking labor that is traditionally unpaid or underpaid.

As the welfare state is increasingly dismantled and weakened, the health of the free market system relies more and more on nonmarket caring labor performed by families and communities. As women's power grows, many are unwilling to accept the unfair bargain of shouldering a double workday. Those families that can afford to escape the increased burden of care do so by displacing these responsibilities down the global care chain, relying on the country's most vulnerable people—women of color, immigrant women, and poor women—to pick up the slack.

In 1999, the United Nations Development Programme argued that a democratic response to the costs of unfairly sharing care requires a renegotiation of individual rights and social obligations by striking a new balance between family, state and market to cover the costs and share the responsibilities of care (UNDP 1999). The UNDP suggests a number of ways we can begin to strike this new balance. We must begin by acknowledging that care is a human priority, though its outcomes are difficult to quantify and its impacts are difficult to capture. We must challenge gendered social norms that unfairly burden women, creating a culture of "universal caregivers" (Folbre 2001) to share social reproduction labor more fairly. We must create incentives and rewards for care—both paid and unpaid—to increase its supply and quality, and remove the "care penalty" in current policy that impoverishes many women. We should increase the supply of state-supported care, aggressively supporting successful programs like Head Start and building toward providing a national system of free or low-cost child care. We might also institute a parenting stipend or tax credits for care work to decrease the risk that people take when they choose to care for others.

Box 8.2
WYMSM Member Profile, Patty Marshall

"I used to be shy when I first came to the YWCA!"

Patty Marshall was born in Albany, raised in Rensselaer, and moved to the YWCA of Troy-Cohoes in August 2000. A small-statured, quiet white woman, Patty loved children deeply and worked for many years as a teacher's assistant in the YWCA's day-care program. She was very active in her church, Sand Lake Baptist, participating in weekly Bible study and vacation Bible school.

She became involved in WYMSM because she was interested in technology and community organizing. "WYMSM impacted people," she told me in a 2004 interview, "especially events like Pavilion. People really enjoyed that. I enjoyed it, too: getting out in the community, talking to different people. I liked doing that—getting to know what other people do, how they feel. I didn't really do that before WYMSM. I used to be shy when I first came to the YWCA!"

Because she was shy, people tended to underestimate Patty, but she was always paying attention and storing information for later use. She was quick to remind you of a responsibility you had overlooked or a promise that you made and had not yet kept. She was very responsible and steady—she never missed a meeting. She also had a wicked, if quiet, sense of humor. She was a behind-the-scenes powerhouse at the YW, working in housekeeping, the kitchen, and at the front desk, and the YWCA presented her with its Woman of Inspiration Award in 2008.

Patty continued her political work after WYMSM disbanded and became a member of Our Knowledge, Our Power: Surviving Welfare in 2007. She passed away in November 2008, at the age of forty-six, after a short illness. She is sincerely missed.

Drawn from an interview on April 6, 2004, and field notes, 2001–2004

Finally, single parents should not be required to work outside the home when their children are young. It is bad economic policy, as it costs the federal and state governments much more to pay for child care while single parents seek employment than it does to provide parenting subsidies. The welfare reforms of 1996 made finding paid employment the legal obligation of single parents, negatively impacting their children and repudiating their worth as caregivers (Mink 1998, 103). Though the burden of combining parenting with the expanded work requirements falls mostly on women, single fathers are increasingly feeling its negative impacts as well. If the dual economy is a volatile economy, we must provide destigmatized, adequate support for those who bear an unequal share of the burdens of its risk and vulnerability. To create a truly just information age, we must depenalize care work and care about caregivers.

4. Raise the Floor

The recent increases in the federal minimum wage will do much to help women in the YWCA community. In 2007, after being stuck for ten years at $5.15 an hour, the minimum wage was raised to $5.85. In 2008 it was raised again to $6.55, and in November 2009 it was raised a final time, to $7.25 an hour. Under the 2006 minimum wage, an individual working full-time, fifty weeks a year, made only $10,300 before taxes, which put her a mere $500 above the 2006 federal poverty line for a single individual and left her deeply impoverished if she had children. In 2009 the same worker made $14,500 per year before taxes, $3,670 above the 2009 federal poverty line for a single individual. However, if that worker has a child and is the only income provider in the household, her family will remain $70 below the poverty line. A family with two children and two full-time earners, both working at minimum-wage jobs, will make $29,000 a year before taxes, too much to qualify for federal assistance available to those below the poverty line ($22,050 for a family of four) but too little to be able to afford health care, child care, or savings for education or emergencies. If growth in the low-wage service sector continues to be a primary feature of the information economy, too many full-time workers will remain working poor.

As Annette Bernhardt and Christine Owens argued in their 2009 *Nation* article, "Rebuilding a Good Jobs Economy," we are presented with a unique opportunity in the current global financial crisis. They argue that deep and growing inequality is the biggest challenge for America's economic recovery: while a handful of people prosper and workers are more productive than ever, a decreasing share of corporate profits goes to wages, and benefits are shrinking. Americans are working harder and working longer hours, but wage inequality keeps many families on the edge of economic disaster. Bernhardt and Owens argue that economic recovery should not focus solely on creating jobs but on building a sustainable and just "good jobs economy." To do this, they suggest four strategic policy initiatives: fully enforce minimum-wage and overtime laws; harness government spending to create living-wage jobs; raise the minimum wage even further to stimulate growth; and enact the employee free choice act, which would guarantee workers protection from intimidation and harassment should they choose to unionize.

Raising the minimum wage and building a good jobs economy should do much to raise the floor for America's poorest workers. But for those unable to participate in the labor market full-time because of caretaking responsibilities, we must expand and depenalize public assistance. The crucial work of caretaking—raising children, caring for elders, tending the sick and infirm—should not be the fast track to poverty, stigma, and political disenfranchisement. Welfare should be expanded to truly meet the needs of poor and working-class parents, to help them acquire education, find good jobs, and save for their family's future.

5. Revive a Vibrant Democratic Culture and Expand Cognitive Justice
One of the key goals of popular technology is to foster more critical, more inclusive thinking about the relationship among science, technology, social justice, and citizenship. Science and technology issues and debates provide excellent opportunities to enrich democratic culture and expand cognitive justice. There are many intriguing models for institutions that help create more critical technological citizens. Two of my favorites are consensus conferences and science shops.

Consensus Conferences   A consensus conference is a process of deliberative inquiry that gathers together groups of "nonexpert" citizens to study and deliberate on a specific scientific, technological or social problem and make recommendations about future actions. The process is in use across the globe—Argentina, Canada, India, Japan, Norway, the United States, and Zimbabwe are among the countries that have held consensus conferences—though the process as it is currently practiced was developed in Denmark. A focus on deliberate inquiry is central to consensus conferences, and they share some common elements: (1) panelists are individuals who do not have a direct stake in the issue being discussed but have an abiding interest in the problem as taxpayers and citizens; (2) the diversity of the citizen panel participants reflects that of the larger population; (3) the deliberative process is informed, ongoing, and facilitated to identify points of consensus.

The Danish example is particularly instructive.[1] After the Danish Board of Technologies (DBT) formulates a question to be addressed by a consensus conference, they send 2,000 invitations to a random sample of Danish citizens. Of the responses, which include a brief statement of background and the respondent's motivations for involvement, participants are chosen to reflect the age, gender, educational, occupational, and geographic diversity of Denmark. The group gathers for several informal meetings to receive background information on the question under investigation, formulate an approach to the topic, and draft questions for the conference itself. The DBT provides a professional facilitator and an advisory/planning committee to aid participants throughout this process. The conference begins when a variety of experts are called to make presentations outlining their understanding and recommendations for the issue at hand. The citizen panel cross-examines the experts based on the questions and concerns they developed in earlier meetings. At the end of the conference, the citizen panel spends several days deliberating, developing points of consensus (bearing in mind that there will always be some disagreement), and preparing a report outlining the issues that bear on the topic and their recommendations. The panel then makes a public presentation of its findings and the resulting report to government officials, policymakers, other citizens, and the media.

Science Shops   Science shops[2] are "university-based organizations that do *pro bono* or low-cost research for community groups . . . solicit[ing] questions from community groups and then find[ing] university researchers to conduct research in the natural, technical, or social sciences" (Farkas 2002, 3). Science shops arose out of Dutch student movements for social justice in the 1960s and 1970s and aim to reach both from science out to the community at large, and from the community into the practices of science. Science shops are basically matchmaking institutions that mediate between community organizations, which have specific questions and needs, and university students and faculty members, who have research agendas to fill. Science shops play an important translational role, helping community members frame problems in ways that are amenable to academic research and helping academic researchers shape their agendas to fit real-world problems and concrete community needs. Science shops aspire to more widely distribute the knowledge production capacities of universities, provide civil society with usable knowledge, and increase public access to science and technology. In doing so, they create institutions for producing equitable and supportive research partnerships, enhance understandings among policymakers and universities of the research and education needs of broader society, and enhance the skills and knowledge of all partners in research projects.

In the United States, the Loka Institute (http://www.loka.org) stands out as an exemplary organization attempting to kindle popular participation in decision making about science and technology. Formed in 1996 in Amherst, Massachusetts, Loka holds conferences and produces publications about community-based research; organizes citizens' panels on telecommunications policy, genetic testing, and nanotechnology; connects science shops in Central and Eastern Europe; and tracks consensus conferences on science and technology policy worldwide. The DataCenter (http://www.datacenter.org) in Oakland, California, also stands out as an exceptional model of democratized research and decision making. The organization provides research support and technical assistance for social movements undertaking their own research agendas. They are vocal and tireless advocates for creating a culture of "research justice," where all communities are "able to reclaim, own and wield ALL forms of knowledge and information as political leverage in *their* hands to advance their own change agendas" (DataCenter 2007, 1).

6. Spread It Around

There are certainly pressing reasons to follow distributive paths along the road to high-tech equity. Though women in the YWCA community had a great deal of interaction with technology in their everyday lives, they did not always have the resources and information they needed to become active producers of technology and media. Though access-based solutions to inequality, such as community technology centers (CTCs), can solve only a few of the important high-tech equity concerns I address in this book, they nevertheless fulfill important community needs and may act as "centers of gravity" for citizen engagement and political organizing. CTCs can provide access to IT, deliver support and training, and help clients create community content. Many CTCs are in locations already dedicated to community building and providing social services, connecting the goals of technology access and community building, and explicitly linking technological goals to social movements. Community technology centers may be freestanding or located in libraries, community centers, and public housing, adding value to the public services offered at those sites.

I have been honored to work with some great CTCs in my time, among them Plugged In in East Palo Alto (http://www.pluggedin.org). Plugged In faced serious struggles, including having to lay off most of its staff, after being removed from its original Whiskey Gulch home. The organization fell victim to many forces that converged in the mid-2000s: neighborhood gentrification, the waning of high-tech philanthropy, and the erosion of federal digital equity programs under the Bush administration. Another organization that I much admired was the Low Income Networking and Communications Project (LINC). Rather than making poor and working-class people come to a central site for technology access and training, LINC, a project of the National Center for Law and Economic Justice, sent mobile technology "circuit riders" to help grassroots organizations build technological capacity. The project lasted for eight years, ending in August 2006.[3]

Too many innovative, nationally recognized community technology organizations folded as political sympathies and philanthropic priorities shifted in the mid-2000s. We should recommit to federal programs for building technological access that have languished over the last decade: the Technological Opportunities Program, the Community Technology Center Program, and the Department of Housing and Urban Development's Neighborhood Networks Program, for example. These programs help technology access points build infrastructure, create community-centered programming, and provide continuity. If you are interested in high-tech equity and not sure how to contribute, volunteering in a CTC is a great way to start.

7. Protect Our Rights to the City

The information economy is not placeless. It is taking place in cities, towns, and communities like yours and mine across the country and across the world. Place matters; we should not think of our geography as primarily mutable. Though cosmopolitanism has its benefits, it is not equally available to all people, and it cannot fully replace a sense of home. There are many fine policies and institutions that exist to stop gentrification, create and maintain livable communities, secure the benefits of economic change for the communities where development is undertaken, and protect the cultural rights of existing residents in areas undergoing rapid change.

Only the District of Columbia and a few cities in four states—California, Maryland, New York, and New Jersey—have rent control or stabilization laws. These laws limit how much and when rent can be raised and create a rent control board that, among other things, decides the maximum amount a landlord can charge for rental units and conditions under which a tenant can be evicted. Rent control and rent stabilization protect individuals and communities in times of rapid economic change, guarding long-time and low-income residents against displacement during speculative real estate booms. Most small cities, like Troy, lack rent control, and are thus especially vulnerable to rapid gentrification.

Like rent control, community benefits agreements (CBAs) protect existing residents in times of volatile economic change. CBAs are legally enforceable contracts signed by community groups and developers that set out a range of benefits communities can expect for hosting and providing resources to a private development project. These benefits may include securing grocery stores, funding job training or providing apprenticeships, developing youth centers, targeting workforce outreach to minorities and other disadvantaged groups, constructing affordable housing, meeting living-wage requirements, or a wide range of other benefits negotiated between community representatives and developers.[4]

Public housing is a crucial institution for promoting family and community health that is increasingly under attack. Like the city of Troy, too many communities are allowing public and affordable housing to deteriorate. Once blamed for the decline of the American city, public housing is now being forced to the edges of American urban space, as middle-class individuals and families return to small cities across the nation. In Troy, we have seen a slow but steady move of public housing out of downtown and into the suburbs, where residents have less access to community resources, public transportation, and social networks that can provide resources for political action and mutual support. Increasingly geographically and socially marginalized, public housing residents bear more than their fair share of the burdens of community change, being forced to travel farther for work and school, lacking effective and adequate community services, and living in desolate and stigmatized locations. Everyone deserves safe and affordable housing, and public housing must be protected as one of few alternatives for poor and working-class families, the elderly, and the disabled.

Gentrification and poorly planned growth do not merely threaten economic rights, they threaten human, cultural, and political rights as well. The attempt to urbanize human rights in the United States, demonstrated by members of the Right to the City Alliance and other organizations, recognizes that issues like transportation, housing, and education in urban centers are linked, locally and globally, to economic justice, democratic participation, cultural protection and expression, and the right to public space for marginalized groups—women, people of color, lesbians, gays, bisexuals, the transgendered, the working class, the poor, and indigenous people (Perera 2008). As David Harvey has persuasively argued,

> The question of what kind of city we want cannot be divorced from that of what kind of social ties, relationship to nature, lifestyles, technologies and aesthetic values we desire. The right to the city is far more than the individual liberty to access urban resources: it is a right to change ourselves by changing the city. (Harvey 2008, 23)

The right to the city movement provides an important challenge to high-tech development boosterism and neoliberal governance, connecting an analysis of the rise of the two-tier information economy and gentrification with a commitment to building collective political power among dispossessed communities in global cities.

In my hometown, ARISE (A Regional Initiative Supporting Empowerment) stands out as an organization undertaking the hard day-to-day work of building just communities in the information age. This loosely affiliated group boasts more than 12,000 members from a mix of faith-based and community organizations. ARISE has task forces on youth and education, criminal justice, regional renewal, workforce development, civil rights of immigrants, and voting. Its comprehensive and thoughtful "equity agenda"[5] helped me imagine high-tech equity as encompassing a wide variety of citizenship, political, and economic rights. The group is currently negotiating the first CBA in the Capital Region, and has worked tirelessly in favor of affordable housing and just regional growth.[6]

8. Clean Up after Yourselves

Despite its clean image, the high-tech economy, particularly high-tech manufacturing, has proved extremely toxic to the natural environment and dangerous to human health. Municipal and business strategies that aggressively court high-tech industries often result in regional "ecocide."[7] According to the Silicon Valley Toxics Coalition, Superfund sites in Silicon Valley are disproportionately located in communities of color, immigrant communities, and poor communities, which continue to suffer the costs of high-tech legacy pollution after manufacturing moves to areas with less citizen resistance, weaker environmental regulations, and fewer labor protections. Low-paid service workers in high-tech areas like Silicon Valley or Silicon Gulch have little choice but to live in the most environmentally degraded neighborhoods, travel extremely long distances for work, crowd too many people into substandard living spaces, or a combination of all three. Environmental racism and injustice force workers to trade their family's long-term health for their immediate survival.

High-tech devices have their most toxic impacts at the beginning of their life cycle, when they are being produced, and at the end, when they are disposed of or recycled. Currently, the United States exports most of its high-tech waste to countries that have weaker environmental protection laws—like Ghana, India, and China—where it is dumped, burned, or dismantled and mined for precious metals and compounds, a process somewhat cynically known as "electronics recycling." The United States exports more hazardous electronics waste than any other country in the world (Stephenson 2008). The recycling process can be extremely dangerous in the absence of proper worker and environmental protections. Workers dismantle computers—often with their bare hands, over open fires or boiling pots—to reclaim precious materials, exposing themselves and their families to lead, cadmium, mercury, hexavalent chromium, and other highly toxic compounds (Puckett et al. 2002, 9). The globalization of high-tech manufacturing and the export of high-tech waste further complicate issues of environmental justice by moving hazardous wastes from rich communities to poor communities throughout the world.

One solution to these monumental problems is to increase manufacturer responsibility for the downstream impacts of their products. Environmental stewardship and green business programs ask manufacturers to eliminate hazardous materials or design for disassembly, and put pressure on governments to create policies that hold manufacturers responsible for end-of-life management of products. Some computer producers, including Apple, are instituting computer take-back campaigns that put the responsibility for disposing of end-of-life computer products (which are classified as hazardous waste) on producers rather than consumers. There have also been some successes in setting higher national standards for the safe disposal of high-tech waste. Fifteen countries in the European Union have signed on to the Basel ban on the export of hazardous waste to developing countries (ibid., 42). Being ultimately responsible for disposing of their own waste will undoubtedly provide much-needed incentives for high-tech manufacturers and countries of origin to design less toxic, more repairable, more recyclable electronic devices.

9. No Sacrificial Girls
Our families, our paychecks, our cities, our drinking water: high-tech equity touches on nearly every aspect of our lives. The ongoing global financial crisis has showed us that the risks and volatilities of the new economy, combined with neoliberal, laissez-faire approaches to governance, can unleash devastation that touches us all. Rather than delivering

a rising tide, the high-tech economy has left many American workers, citizens, and families to drown.

There are no sacrificial people, no *objects* of knowledge, no inevitable victims of progress in this struggle. Every man, woman, and child is equally and infinitely precious. We all have a stake in the creation of a more just information age. Everyone has specific expertise, too, their own experience of the real world of information technology, rooted in their social location. Each one of us has a role to play in creating better knowledge. We cannot afford to neglect any resource for change; there is simply too much work to be done.

If we don't want the information age to deliver widespread economic and political destruction, we must build levees that will hold, craft effective escape plans, and commit to including *all* citizens in a dialog about building a just and equitable information age. For this job we need to marshal all of our strengths, all of our resources, all of our knowledge as a diverse and vibrant country. We cannot put blind faith in science and technology to redeem the American economy and enrich our democracy. This magical thinking has failed us too many times, in ways far too predictable. To create an information age that works for all people, we need to think more clearly, communicate more honestly, and act with more courage and foresight. In the end, our liberation is bound up in each other; we all sink or swim together.

Appendix A: Research Methodology

In the spring of 2004, I traveled to New Market, Tennessee, to visit the ancestral home of popular education in the United States. I stayed at the Highlander Research and Education Center for ten days, enjoying long meals, sitting around evening fires talking and telling stories with southern activists, riffling the archives for what insights I could glean about developing popular technology programs. Though I had many memorable moments at Highlander, the one that stands out is the first time I stepped into the Harry Lasker Memorial Library.

The library is a modest redwood structure overlooking a rolling green field, a cow pasture, and, further off across Route 25W, the Blue Ridge Mountains. Its collection is likewise modest—approximately 6,000 volumes—and its archives are contained in roughly 200 document boxes and five or six filing cabinets full of popular education and workshop materials.[1] Nevertheless, it seemed to me I had finally hit the mother lode: a whole library exclusively dedicated to mutual education! Though I had been exploring participatory approaches to research and education for years, I had only found a handful of volumes, most Freirian or based in the experiences of activists in the global South—Brazil, South Africa, India. But here in the Lasker Library was proof. People's education has a long and vibrant history in the United States. Though I recognize now that my sense of isolation was naive—there is, of course, a large and widely available literature on popular education, citizenship schools, settlement houses, and other people's education projects in the United States and Canada—the validation I experienced was so strong that I stood in the dusty redwood building and wept.

Can ordinary people be smart about something as complicated as the global knowledge economy? Neoliberalism? Government devolution? I think we can. But uncovering that knowledge and systematizing it takes a reorganization of many of the principles of academic disciplines. Participatory research approaches are non-programmatic and highly context dependent. Participatory techniques require enormous practical and theoretical sophistication. Community-based research findings must have high immediate relevance and usability. The focus on diversity and democratic process intrinsic to participatory research is time-consuming relational work that is undervalued in academic institutions, drawing on skills and abilities that many researchers have little incentive to develop.

Traditional academic disciplines would have to shift considerably to fully embrace participatory research. One way to start would be to create a research culture that welcomes a flexible approach to methodology. Methodological purism is not useful for projects that exist in the real world, because real-world problems do not fit into the neat categories recognized by academic disciplines. All methods have benefits and drawbacks; all research techniques should be evaluated in the moment for what they bring to the solution of a specific problem, in the context of a particular place, in the hands and minds of unique people. Because so many WYMSM projects responded to local conditions, included multiple researchers from a variety of social locations, and required immediate relevance to community problems and situations, the research methods that inform this book are many, varied, and drawn from across the social sciences and humanities.

The research methods I used included interviews; composite stories; quantitative analysis of census, Bureau of Economic Analysis, and Bureau of Labor Statistics data; textual analysis; collaborative discussion and reflection; focus groups; and even some plain old investigative journalism. Many—though not all—of the research projects were approached from a participatory action research orientation and included the input of women in the YWCA community in all phases of research design: identifying research questions, choosing appropriate methods, collecting and analyzing data, and deciding on the format and use of the research results. Much of the research was undertaken in collective forums, including large public meetings and workshops, more focused community meetings, and WYMSM meetings.

Space prevents me from detailing every research and action project undertaken at the YWCA (see appendix D for a comprehensive list of popular technology projects). I describe many of our methods—WYMSM's interviews and composite story writing, for example—in the main body of the book. What follows is a summary of data gathering and analysis methods not described in detail elsewhere.

Summary of Data Gathering

WYMSM Meetings

At the heart of this book is the collaborative work of Women at the YWCA Making Social Movement (WYMSM). We held forty-six meetings, each lasting approximately three hours, between November 12, 2001, and July 16, 2003. After the first few meetings, agendas for WYMSM activities were collaboratively developed by group members. We produced minutes for each of the meetings, and also recorded them on audiotape. Because the tapes proved extremely difficult to transcribe—each tape included several speakers, and the conversation was usually intense and lively, with voices overlapping—I transcribed only eight of them, chosen at random to represent each of four periods of five or six months. In addition to these transcripts, I gathered and coded approximately 700 pages of agendas, minutes, notes, flipcharts, game design plans, handouts, promotional materials, and other WYMSM working documents.

Interviews

To supplement data gathered through collective and participatory practices, I undertook twenty-nine semistructured interviews with YWCA residents, staff, and other community members. These interviews provided the opportunity for women in the YWCA community to speak as individuals, with the protection of anonymity (if they chose it), and created a conversational space for in-depth probing of thoughts, ideas, and feelings. The shortest of these interviews lasted just over forty minutes; the longest took nearly three hours. All surviving WYMSM members were interviewed. I did not want to talk to only WYMSM members, however, because the group was clearly self-selecting and not representative of the YWCA community at large. So, I balanced the sample, recruiting seven YWCA residents who had not participated in WYMSM by placing flyers on the halls and in the elevators at the YWCA. I interviewed the housing staff of the YWCA, its human potential advocates, its executive director, and the director of the Sally Catlin Resource Center. I also interviewed four nonresidential YWCA community members who had been active in WYMSM and other activities at the YWCA, as well as one faculty member and two students from Rensselaer Polytechnic Institute who had initiated the collaboration between the YWCA and the university back in 2000. I transcribed the interviews and followed the data analysis procedures described in the section "Summary of Data Analysis," below.

The interviews took place between July 2003 and April 2004. By the time I began them, I had been working in the YWCA community for close to two years. This accounts for the familiar, conversational tone of many of the interviews. I have tried to represent the character of the interviews by including my questions, comments, and prompts in any long excerpt from transcripts. Many of the people I interviewed knew me well and had often worked on projects with WYMSM or in the tech lab. The interviews were but a snapshot—one frozen moment in a very long set of conversations—intended to prompt more in-depth reflection about the activities we were engaged in, to inspire new insight into the information age, and to evaluate our process in collective projects. All interviewees received a copy of the transcript of their interview and approved its use.

There was substantial time for more informal talk, following the threads suggested by interviewees, but I had lists of interview questions to prompt discussion (see boxes A.1 and A.2).

Collaborative Action and Reflection

I spent approximately 1,500 hours engaged in collaborative social change and research activities with women in the YWCA community in a variety of situations: at large public events, in one-on-one conversation, while working in the community technology lab and the Sally Catlin Resource Center, at community meals, standing outside on the porch, running into YWCA community members at the laundromat and corner store, having them over to my house. From these interactions, I produced 372 pages of typed, single-spaced field notes in two formats, a daily log and open notes.

Box A.1

Everyday Experiences with Technology

What kind of technology do you use or come into contact with in your daily life?

Tell me a story about an experience you've had with technology which made you feel really empowered or hopeful.

Tell me a story about an experience you've had with technology which made you feel really disempowered or angry or freaked out.

Would you finish the following sentence? "A computer is like a. . . ."

Technology and Inequality

A lot of people who write about computers say that the reason that economically disempowered women don't engage much with technology is because they're afraid of it, or they just don't understand how important it is. What do you think?

People also say that there is a "digital divide" in this country: that we can all be split up into technology "haves" and technology "have-nots." What do you think? If that's the case, what would your solution be? If that's not the case, how would you describe the situation better?

If tomorrow, all technology was suddenly being used primarily for social and economic justice, what would it be doing? How would the technology itself look different? How would the world change?

Knowledge and Information

What kind of knowledge and information do you have that makes a powerful difference in your life?

What knowledge or information don't you have that, if you had it, would radically transform your life? Our community? The world?

Hopes and Dreams

The Capital Region is dedicating itself to "high-tech" economic growth. If you had the ear of the mayor of Troy or of Albany, how would you advise him to make high-tech growth socially just, fair, and equal?

The YWCA is trying to develop technology programs in the service of social justice. What advice do you have for us? Have you come to any of our stuff so far? Why or why not? What did you think? How will we know when we're doing it right?

Box A.2

Additional Questions for WYMSM Members
What led you to be involved with WYMSM?
What do you think of our mission? Do you think technology can be used as a tool of social change? How?
What was WYMSM's biggest impact? On the community? On you personally?
The "technology" aspect seems to have fallen out of many of the projects. Do you feel like it did? Why do you think that happened? Did it affect your desire to be involved at all, one way or the other?
How effective and equal do you think our process—how we went about setting goals and making decisions—was? What do you think could be changed to make our process better?
What role do you see for organizations like the YWCA in supporting WYMSM's work to "use technology as a tool for social justice"? How can the YW best support our work?
What role do you see for universities (such as RPI) in supporting WYMSM's work to "use technology as a tool for social justice"? How can RPI best support community-based activism and research?
Do you consider yourself "political"? An activist? A community organizer/ worker? Would you like to be any of these things? If so, what keeps you from doing this work more effectively?
Where do you see WYMSM going in the future?
In a perfect world, and if we had all the resources we need, what would WYMSM be doing in ten years?
What three things can we do now to help achieve that goal?

The daily log asked me to reflect on what I observed and learned in the community each day and what kinds of activities I was engaged in, and provided me space to interpret successes, failures, interpersonal interactions, and my emotional responses. I found the log offered much-needed structure and saved me from "blank page syndrome" when I got home after a long or emotionally charged day at the YWCA. When I was reporting on specific interactions, such as staff meetings or community meals, or on large public activities, such as the Hunger Awareness Day event or workshops, I wrote open notes, describing what happened, my reactions, and any insights gained as soon as possible after the event had ended (usually that night, but in some cases the following morning). I've reproduced one daily log in box A.3.

I hesitate to characterize this work as ethnographic participant observation. Though ethnography is a remarkably useful method for providing rich descriptions of social worlds, institutions, and actors, I have found that the terms "ethnography"

Box A.3
Daily Log

YWCA of Troy-Cohoes
April 10, 2003

List things that happened today:
Jes C. did drop-in hours in the tech lab from 12 to 2, and made a flyer advertising a field trip to WRPI, the local college radio station.
Christine N. produced the tech lab weekly calendar and sign-up sheets for next week's class.

List things you did today:
Checked messages and email.
Sent notes about WYMSM meeting to members.
Opened for [the volunteer tech lab instructor].
Went to community meal.
Held drop-in hours in the tech lab.
Lunch with Christine N. to catch up. See additional notes dated 4/10. Set new priorities for post-break work.
Before work, I ran into [YWCA resident] at the Laundromat on 4th Street, and we talked about the war. She supports it—thinks that Hussein is a dictator and the U.S. is liberating Iraqis. She has two nephews in the military currently. It was a good conversation, and we ended up talking about how poor folks—on both sides—are always victims. She said, "If the money [from oil, for reconstruction] doesn't go to the Iraqi people after the war, then I'll become a protestor!" Her mom was a civil rights protestor and she says she is "essentially nonviolent."

Describe an interaction a community member had with technology today.
[YWCA resident] and I sat down during drop-in hours to learn HTML. She got it very quickly, the syntax and everything—so much so that she was jumping three steps ahead and making very good guesses at what tag names were going to be. But her big lightning bolt moment came when she got the connection between what we were writing in Notepad and what was displayed in Internet Explorer. There's something there about impact, I think, moving from private (word processing) to public (Web browser), or about linking language and action in Tardieu's sense.

Describe an important or evocative social interaction that happened today.
Had a conversation with [the volunteer tech lab instructor] while we were waiting for people to show up for her class. We were talking about the war and stuff. She's very concerned with economic impacts at home. The conversation started because I mentioned how difficult it currently is to find grants. She said that she had never been laid off before 9/11, and since then

Box A.3
(continued)

she has been laid off twice—both technical jobs, one at a start-up (she called it an "Upstart"!!).

Describe a resource connection you made today.
[Troy community member] has potentially offered to pay Jes to teach/mentor technology stuff to [community program] participants. It would be super-great, I think, if we could use their presence in the lab and support of Jes to break down the "us and them" border between the YWCA and the rest of the community by mixing [program] participants and YWCA residents in the classes. I think it would be beneficial for both organizations.

Describe a "life-critical information need" you observed or experienced today.
[YWCA resident] in the lab uses MSN chat to stay in touch with her family—also uses a second email account as a sort of photo album with pictures of her family, friends, friends' children, etc., that can't be lost or misplaced if she has to move in a hurry.

Describe one of today's successes.
When I dropped in to visit yesterday, Christine N. said, "We're glad to see you back, but we're really glad you left." In my absence [for a week's vacation], Christine, Jes and the advocates were forced to really figure out how the YWCA can support the tech lab (and vice versa). Christine N. met with [staff members] and they discussed having staff present at drop-in hours a couple of times a week. Also, they found a van for the field trip to the radio station, and people seem really excited about it. I am really excited about the shift and so are they—I think they were feeling a little excluded before. The important lesson here, I think, is about the deadening effect of "technological expertise." When we see the lab as primarily a technological resource and as my job, I think both staff and residents feel excluded from it. When we see the lab as a *community* resource—the responsibility of the staff and the rest of the YWCA community—it becomes more useful, more inclusive, more connected to the YWCA's mission, and—I think—it will be used more and better.

Describe one of today's failures. What fell through the cracks?
Nothing today!

What surprised you the most?
People are increasingly seeing me as tech support. [YWCA resident] yesterday asked me to figure out how to get her laptop online. [Another resident] really insisted that I come up to her room right away and work out a conflict between her online service and her answering machine. I managed to put her off until Saturday morning, but . . . !

Box A.3
(continued)

How do you feel?

Awesome. Like we're on exactly the right path, both for the YW and for me. I'm even finding time to write and read. And eat well!

---

or "participant observation" often imply to readers divisions—between observer and actor, insider and outsider, subject and object of knowledge—that were neither desired nor possible during this project. I recognize and take seriously the many and meaningful power relationships within the research project.[2] However, I use the phrase "collaborative action and reflection" instead of "participant observation" to represent the truly conversational nature of the research.

Summary of Data Analysis

Following a grounded theory approach (Glaser 1998; Glaser and Strauss 1967), I coded interviews and WYMSM meeting transcripts, print documents, field notes, meeting agendas and minutes, flip charts from meetings and workshops, and YWCA community ephemera, identifying keywords that arose in each document. There were literally hundreds of keywords, from the relatively mundane "EBT card," "hours the lab is open," "IT bubble," "medical technology," "public access," and "support networks" to the incredibly complicated "dependency," "displacement," "democracy," "monitoring," "politics," and "prejudice." As my data set grew, I compared documents with each other, and these keywords started to cluster together, pointing to broader themes and suggesting conceptual relationships. Important themes that emerged included access, citizenship, poverty, power, process, and "the System." Some of these themes were surprises to me: I had little or no interest in the social service system when this project began and had no intention to write or think deeply about it, for example. Other themes, such as access, I was forced to consider in an entirely new way.

Because I remain in the YWCA community, I was able to check in with others, especially WYMSM members, throughout the process of writing this book to confirm my interpretations of the data or clarify points I did not understand. I continued to discuss the book and negotiate shared meanings with WYMSM members until the final days of manuscript preparation. I practiced a kind on ongoing, negotiated consent with my collaborators, offering them the opportunity to clarify their points, change their anonymity status, and control the form their ideas and the publication took.[3] I believe this increased both the ethicality of the research and its validity.

As my set of keywords and themes grew, I began to write notes to myself about how the concepts that were emerging from the data matched up—or did not—with existing understandings and frameworks in the literature about technology and poverty in informatics, political science, social studies of science and technology, and women's studies. Over time, the thesis of the book grew out of this mountain of data. So did the structure. Chapters 3 through 5 each deal specifically with an aspect of the relationship between technology and women's poverty that was identified as central to social justice in an age of IT by the women in the YWCA community: policy and technology access; economic development and the high-tech, low-wage workplace; and welfare administration technologies. Though the data set was massive, most of my coding and analysis was done manually by writing in the margins of transcripts, using multicolored highlighters to group themes together, and creating conceptual index cards that I could move around to help identify and explore relationships within the data. Near the end of my writing process, I began using the qualitative data analysis software ATLAS.ti to make sure there weren't relationships or insights in the data that I had missed.

Appendix B: WYMSM Sample Agendas

The following WYMSM meeting agendas and minutes are provided to give the reader a sense of the actual activities undertaken on a week-to-week basis by WYMSM. I have provided four agendas, for meetings five or six months apart, to show how the group's concerns and process changed over time.

Agendas and Supporting Materials

Workshop 1
Sally Catlin Women's Resource Center
Troy-Cohoes YWCA

November 12, 2001
3 p.m.–8 p.m.

Theme: Creating our own Game of Life
For Whom: Residents of the YWCA
How many? 20
How long? 5 hours

Objectives:
• Present the concept, relevance, and some examples of simulation and its effect on participants' lives
• Identify how simulation could be useful to participants and explore how existing examples succeed or fail to do so
• Identify four to eight participants for ongoing design team project and technology training

1. Introductions and Group Facilitation (60 minutes)
This first exercise (or several exercises) should introduce us to each other, open up space for participants to feel comfortable talking, and begin to identify common problems, issues, barriers, concerns, benefits, and promise that the group works within or toward (begin to sketch these out on the flip chart; we'll come back to this in step 7).

- Introductions: Who we are, why we're here (5–10 minutes)
- Icebreaker: Memorable moment (15–20 minutes)
  Have each person tell the most memorable moment in their lives so far, and their most memorable moment at the YWCA so far. [Facilitator: Jes]
- Exercise: A Fantasy of a Life Without Women (20–30 minutes)
  In groups of four (break into five groups by putting numbers under seats), ask the women to imagine and describe: "What would happen in your family, your neighborhood, and your town if all the women went away for one month?" After about 15 minutes, share informally with the whole group. [Facilitators: All—one in each group, led by Pat]

2. Sketching Process and Structure (10 minutes)
Coordinators and participants sketch out (1) how the workshop will be structured, (2) what we all hope to gain from the process, and (3) what content we hope to cover on the flip chart. We should also have a discussion at this point about group norms and ground rules. [Facilitators: Virginia and Christine]

3. What Is Simulation? (20 minutes)
Introduction to the concept of simulation, in three parts:

- Basic description of simulation: developing "best guess" pictures of complex situations (4 minutes) [Virginia]
- Simulations used in social service systems: welfare tracking and forecasting (developing pictures of who we are, models, and policy polling) (8 minutes) [Nancy]
- Participant simulation: SimShoBan: general explanation of the process of developing simulation of Native American technology (include storyboards) (8 minutes) [Ron]

4. Exploring Simulation (90 minutes)
Split into four groups, two exploring the ROWEL simulation and two others exploring The Sims.

A. Getting into groups
Finding group navigators (2 minutes)
First ask how many people want to do each exercise. Once we've established the number of groups to do each simulation, we want to ask for four to six volunteers who would be comfortable navigating through The Sims and reading/distributing the ROWEL activity.
Barnyard Bingo (3 minutes)
This may not be necessary once we've asked people what they want to do. Begin with four to six groups of index cards, each group with the name of an

animal on it. First, we make sure that each of the navigators receives a different animal. Next, we ask residents if there is a specific simulation they want to do. (We keep them in mind when handing out the note cards so that they get one of the animals that the respective leader has.) Then we distribute the rest of the note cards so that everyone who is participating has one. We explain to participants that three or four people in the room have the same animal on the note card, and the goal is to find them by making the sound your animal makes. Once group members have found each other this way, we're ready to start.

B. Exploring *The Sims*

Introduction of activity (10 minutes)

Introduce the concept that simulation can be used both as a tool or a game. For instance, the ROWEL simulation is an educational tool that can be used for policymakers, social workers, and the government. On the other hand, *The Sims* is a game that is popular with kids of all ages. So, given the two different purposes, there will probably be differences in the two simulations.

Ask participants, what would a simulation that was meant for *each of you* be like? (Note: *Each of you* is stressed so that we are not labeling what it means to be a resident at the YW.) Could a simulation meant for each of you be similar to this game or this educational tool?

Exploring the simulation (45 minutes)

This part seemed pretty self-explanatory, although we did think of a few things that might be good to keep in mind:

Making a reference to the group norms will hopefully emphasize letting everyone talk and participate. A few options for structure are a round-robin approach or letting each participant choose a role. Having a paper and pen handy to take notes might make it easier to discuss what they learned, liked, and disliked.

Introducing/discussing the simulations with each other (30 minutes)

After an hour, teams come back together to introduce the games to each other and share what they've found and discussed. Ask a volunteer from each group to write it all down on the flip chart.

Questions for this exercise might include the following: How is this game relevant to your experience? How is it NOT relevant to your experience? What techniques does it use to convey implicit and explicit information? If you were going to remake the game, how would you do it differently?

C. Exploring Life in the State of Poverty

Introduction of activity (5 minutes)

Introduce the concept that simulation can be used both as a tool or a game, but stress that the makers of Life in the State of Poverty try to be very clear that it is a *simulation*, because it's based on real experiences and data. Talk about who makes it (Reform Organization of Welfare, a grassroots Missouri organization of low-income people and their allies working together to inform the larger community about the inadequacies and injustices of the current welfare system), whom it is intended to educate, how it works, and who benefits from it.

The welfare simulation experience is designed to help participants begin to understand what it might be like to be part of a typical low-income family trying to survive from month to month. The objective is to sensitize participants to the realities of life faced by low-income people. The welfare simulation experience can be an eye-opener for many audiences, including professionals who provide services for low-income families, church groups, women's organizations, college students, and politicians. The success of the simulation relies on the participation of "staffers" who have faced or are facing poverty. The dialog between staffers and participants is key to getting the most from the simulation experience.

Explain that the use of this simulation in the workshop is intended to do two things: (1) give an example of simulation for social change and (2) try to identify residents who might be interested in being staffers for a full run of the simulation in the spring.

Choosing roles (7 minutes)

Explain the roles to be filled, both staffers (residents) and families (us). One facilitator (Virginia or Jes) will be the director, whose job it is to answer questions about the simulation, keep time, and pass out Luck of the Draw cards. Divvy up the roles and distribute packets. Allow a few minutes for the facilitator to explain what each person does, if they don't understand.

Goals of the game (3 minutes)

The goals of the game are to keep your home secure, feed your family, keep the utilities on, make necessary loan payments, pay expenses, and meet unexpected situations. Remember: No healthy teenager likes to sit quietly at home. A child who has not eaten all day will cry and complain. Adults seeking work are often frustrated or irritable. Parents can get desperate in their search for food and shelter for their children.

Running the simulation (30 minutes)

We do two fifteen-minute "weeks" of the full hourlong "month" of the simulation.

Evaluation and dialogue (15 minutes)

Families: How many of you think you would have been able to pay the rent, keep the utilities on, buy food, and so on, at the end of another two weeks? What happened to your family? What good things? What bad things? How many families improved their situation? Maintained good relationships? Other insights?

Staffers: Comment on how the families played the game. Were there any surprises? How would you have done things differently? Reflect on how the simulation reflects—or doesn't reflect—your own experiences. What techniques does it use to convey implicit and explicit information? If you were going to remake the simulation, how would you do it differently?

Decide what to present to the larger group and who is going to present it.

Introducing/discussing the simulations with each other (30 minutes)

After an hour, teams come back together to introduce the games to each other and share what they've found out and discussed. Ask a volunteer from the group to write it all down on the flip chart.

5. Dinner at 6 p.m. (45 minutes)

During dinner, coordinators split up to continue the conversation with participants, introduce the idea of the design team we're developing, talk about technology training (what would they like to learn?), and recruit for workshop 2.

6. Facilitating Critique (15 minutes)

What problems could simulation or simulation games help us solve? How would the simulations we've looked at today have to change in order to be helpful in/ reflective of participants' lives? What techniques could we use to create a simulation more reflective of our lives? [Facilitator: Pat]

7. Reimagining Women's Lives, Building Women's Tools (40 minutes)

Introduce framing question. Break up into four groups and use collaging to start to explore what a simulation that could help alleviate problems or attain aspirations might consist of or look like. (30 minutes)

Present finished collages to each other in pairs. (10 minutes) [Facilitator: Pat]

8. Evaluation and Follow-Up (20 minutes)

Methodological walk-through: Have a participant walk us through the stages of the workshop, using the flip chart as an aid. Reflect on how well we met our process and content goals (laid out in step 2). (10 minutes)

Collaborate on next steps, recruit for next workshop, let people know project timetable, that money (in the form of . . .) is available, collaborate on ground rules (which will be written up for the next meeting), and so on. (10 minutes)

WYMSM Agenda for April 27, 2002

Welcome; read WYMSM mission statement

1. Agenda for next week, summer priorities
2. Ideas about field trip to Seneca Falls (after August 18)
3. Letter to YWCA volunteers thanking them for ROWEL
4. Fun activity! (Patty)
5. Creative brief, continued
6. Graphic design, the WYMSM logo, continued
7. The good, the bad, and the ugly (graphic design exercise)—tabled until May 11, 2002.

WYMSM Mission Statement

As an initiative of the YWCA of Troy-Cohoes community, *Women at the YWCA Making Social Movement* (WYMSM) seeks to use technology as a tool of social change.

This community-building collaboration creates projects that help women build awareness of existing resources, knowledge, and experience, precipitate resource sharing and development, and provide supportive encouragement to learn from others' experiences through technological tools and social network building.

WYMSM Minutes for April 27, 2002

In attendance: Jes, Ron, Virginia, Patty, Jenn

Agenda for next week, summer priorities
We started the meeting by co-developing an agenda for next week's meeting and trying to set some priorities for this summer's work. Patty has the agenda we developed. Virginia gave everyone copies of the enterprise grant to look at and comment on.

We decided that, since the self-sufficiency standard calculator will be able to work as most of the first part of the software (decision-making tool), and the enterprise grant may provide for the third part of the software (asset-based database), we should concentrate on getting ready to make the second part of the software (personal and professional development simulation) this summer. Jes will be in a class that can help us design the software itself in the fall, so we need to make sure that all the important pieces get put into place over the summer so we can make best use of those resources.

These include:

• Figure out what technical stuff we have to learn, and learn it.
• Learning Web design, Flash, and Photoshop should be our priorities over the summer.
• *What* kind of data do we need for the simulation? What issues affect women's lives most at the YWCA? How have they dealt with these issues in the past?
• *Where* are we going to get these data?
   Women's Economic Empowerment Series, ROWEL documentation, documentary video work, Jes's Botswana Connection, agencies in Troy, personal folders, the tech lab "Coffee Talks" (starting mid-May).
• *How* are we going to pull data out of these rich experiences (method)? We need to continue with the popular education/participatory action research approach and use it to help us develop a method for finding reliable data. Some ways we might get at existing concerns are active listening (see handout from Training for Transformation), elicitation posters, collaging (expressive arts), going back over existing materials to find out what they have in common, confessionals.

Ideas about field trip to Seneca Falls (after August 18)
We'd definitely like to go to Seneca Falls if we can work it into our schedules. Jes would like us to wait until after she gets back so she can come, too.

Letter to YWCA volunteers thanking them for ROWEL

We finished the last of the thank-you letters; now we need copies of the Record article and a list of Women's Economic Empowerment Series events to include in the envelopes.

Fun activity! (Patty)

Creative brief, continued

We made more decisions about the software, using the creative brief as a tool. We decided on our own goals for the project: finish the software and distribute it, learn technology and computer skills, build leadership skills, learn more about and from each other, build community and social networks, build awareness of important issues and share resources for dealing with them, and have fun!

Graphic design, the WYMSM logo, continued

We worked for a while on the logo, developing twelve ideas for improving it and whittling it down to four choices to continue to develop by voting. Jes will develop two of them and Virginia will develop the other two. We'll bring them back to group on May 11 and make decisions about what parts we want to keep.

The good, the bad, and the ugly (graphic design exercise)

Tabled until May 11, 2002.

WYMSM Agenda for October 9, 2002

Welcome; read WYMSM mission statement

1. Role choice (for next meeting): facilitator, fun activity, minutes taker, vibes, mission reader, minutes reader
2. Fun activity: Jes (10 minutes)
3. Listing assets and skills (15 minutes)
4. Listing interests and topics (15 minutes)
5. Opportunities list: Virginia and Jes (15 minutes)
6. Prioritizing/connecting (15 minutes)
7. Discussion: What connections and possibilities are now visible? Choose projects to follow. (30 minutes)
8. Moving forward: next steps for each project/area chosen (20 minutes)
9. Administrative: extra hour? (tree, HTML lessons, work on a project, research, etc.) (10 minutes)

WYMSM Mission Statement

As an initiative of the YWCA of Troy-Cohoes community, *Women at the YWCA Making Social Movement* (WYMSM) seeks to use technology as a tool of social change. This community-building collaboration creates projects that help women build awareness of existing resources, knowledge, and experience, precipitate resource sharing and development, and provide supportive encouragement to learn from others' experiences through technological tools and social network building.

WYMSM Minutes for October 9, 2002

In attendance: Zianaveva, Ruth, Jes, Nancy, Julia, Jenn, Patty, Virginia
Absent: Cosandra, Christine
First agenda item: Role choice for next meeting: Facilitators: Jes and Julia; fun activity: Ruth; minutes taker: Virginia; vibes and mission reader: Zianaveva; minutes reader: Jenn
Second agenda item: Fun activity—Jes's Name Game
Third agenda item: Listing assets and skills. Brainstorming by all—computer skills: HTML, graphic design, hardware, software, etc.
Facilitating, teaching, public speaking, languages, education
Community-based outreach, diversity, etc.
Fourth agenda item: Listing interests and topics. All agreed to skip.
Fifth agenda item: Opportunities list.

• Verizon grant for women teaching women technology
• Jes's two classes: database class for Community Development Bank and computer games class.
   All discussed how these could help WYMSM develop projects such as board games.
   Pro board game project: Nancy pitched how board games develop cooperative approaches and encourage success. Ruth agrees they develop social skills.
• Virginia announced opportunity with [community member] to take a radio class, create WYMSM PSA and radio show, possibly using ROWEL video as resource.
• City of Troy Enterprise Grant: Pending/no date certain.
• Tech learning opportunities with Virginia: HTML/web; graphic design; other?

Sixth agenda item: Prioritizing/connecting. Virginia asked all to agree regarding establishing goal/task and working toward it. All agreed to two concrete projects to coincide with opportunities presented by Jes's class and money from the COPC grant (and maybe the enterprise grant): the Community Development Bank database and a (computer) game.
Seventh agenda item: Discussion. Jes and Virginia suggested we develop milestones first. All agreed to goals to include: identifying theme, identifying topics, doing research (playing other games for ideas), identifying audience(s), deciding on a name, doing market research, researching comparative examples, storyboarding game scenarios.
Virginia cited the need to shape work according to current and future funding (needs grant research).
All agreed to complete a prototype/pilot by end of year, December 30.

Theme: Ruth identified possible theme in mission statement: "build awareness of existing resources, knowledge, and experience, precipitate resource sharing and development, and provide supportive encouragement to learn from others' experiences through technological tools and social network building."

Discussion of why social change is important: Zianaveva wants wage increases, affordable day care, quality housing. Ruth speaks about quality of life. Jenn says quality of life is the theme, and then we all discuss how.

Eighth agenda item: Next steps and tentative agenda for next meeting. Identify theme of game by doing research (playing games for ideas).

Ninth agenda item: Administrative. Virginia discussed the extra hour to work on tree, HTML lessons, project, research, etc.

Julia, Ruth, and Zianaveva will meet with Virginia Friday. There is funding available to do data entry and mentoring. Jenn will research computer games for the next meeting.

All completed time sheets.

Email Julia with additions or corrections to these minutes.

WYMSM Agenda for March 1, 2003

Welcome; read WYMSM mission statement (Julia); summarize minutes (Jenn)

1. Role choice (for next meeting): facilitator, fun activity, minutes taker, vibes, mission reader, minutes reader, snack (5 minutes)
2. Fun activity: Jes (10 minutes)
3. Interrogating interviews: Read through the last two interview transcripts, code them (30 minutes)
4. Interview conversation (30 minutes)
5. Questions from handout, find keywords, concepts and game content. How can we use this information?
6. Break (15 minutes)
7. Planning for next round of interviews (30 minutes)
8. Reformulate questions for residents/community member interviews
9. Grant writing (30 minutes)
10. Go over possibilities Virginia found. Decide on plan for following up on these opportunities. Prioritize.
11. Video: Nancy (30 minutes)
12. Administrative: timesheets, schedule next several meetings (5 minutes)

WYMSM Mission Statement

As an initiative of the YWCA of Troy-Cohoes community, *Women at the YWCA Making Social Movement* (WYMSM) seeks to use technology as a tool of social change. This community-building collaboration creates projects that help women build awareness of existing resources, knowledge, and experience, precipitate resource sharing and development, and provide supportive encouragement to learn from others' experiences through technological tools and social network building.

WYMSM Minutes for March 1, 2003

We reordered the agenda items to complete nos. 5, 6, and 7. We also set a research schedule for the rest of March.

5. Planning for the next round of interviews (30 minutes)

We spent most of our time talking about how we were going to conduct resident/community member interviews. Nancy suggested a technique of using composite stories to elicit feedback on character profiles.

By combining common elements from the interview data we already have, we should be able to get feedback on creating credible characters in a way that is less invasive, protects interviewee confidentiality, and provides immediate feedback. Common elements = credible characters. The composites will generate game characters and provide a validity check. Interviewees can say whether composites correspond with their experience or do not, and can suggest things we might have missed.

Suggested composites:
Character on DSS and needs to access education. For example, character wants a GED. What is the conflict situation there?
Person who always falls through income cracks (too much to qualify, but not enough to live on).
Person who wants to get married.
Person who wants to get out of a marriage, or is thrown into having to get married.
Persons forced to move back in with their family.
Single parent.
Certification/recertification struggles.
Racial issues.
Disabilities (mental, physical): people who are unable to work and can't gather the resources to get by.
Low-income elder (on SSI).
On the edge of getting off, but needs health insurance.

Suggested themes:
Health insurance, health status (disability), education level, income level, type of job (work history), age, marital status, kids, parental resources, criminal history.

6. Grant writing (30 minutes)

We went over the grant opportunities we have so far, split them up and agreed to do further research on a number of them. We decided to follow up on: Holding Our Own (Patty); Howard and Bush (Cosandra and Virginia); Ford Foundation (Jenn and Jes); Markle Foundation (Julia and Jes); and Rockefeller Foundation (Virginia). We'll check back with each other in the next meeting.

7. Video (30 minutes) (Nancy)

The COPC group at RPI is making a video about the work they've been involved in over the past year. Nancy discussed how to structure our meeting on March 22, when someone from the COPC will come to the meeting to do a video shoot.

Meeting schedule for March

March 8—Interview Analysis, Part II: Develop resident/community member interview questions.
March 15—Pilot new interview questions on each other.
March 22—Analyze data from pilot, reformulate interview questions; video shoot.

Appendix C: Popular Technology Sample Exercises

The information and exercises that follow are intended to provide a sense of what the popular technology workshops included and how they operated. This is only a small sampling of a broad variety of activities: the Women's Economic Empowerment Series alone consisted of nine two-hour workshops, and the documentation of the series fills a two-inch-thick three-ring binder. I have tried to provide a spectrum of different activities, one from the Women's Economic Empowerment Series, one from a large public event held by Women at the YWCA Making Social Movement (WYMSM), and one undertaken by two members after WYMSM disbanded. I have also provided supporting documentation to provide inspiration and offer tools to help readers undertake popular technology projects themselves.

Sample Workshops

Supporting Documentation:

How Much Is Enough? The Self-Sufficiency Standard

Workshop presented August 14, 2002, as part of the Women's Economic Empowerment Series at the YWCA of Troy-Cohoes

Authors: Virginia Eubanks and Christine Nealon

Description
How much does a family need to really get by?

Neither the minimum wage nor current poverty standards take into account geographic concerns or adequately provide for the costs of working. The self-sufficiency standard, developed by Dr. Diana Pierce, is different. Pierce's self-sufficiency standard takes into account the costs of employment, particularly child care and transportation, and the cost of health care. It adjusts for geographic differences in cost, especially housing; can be updated to reflect revised understanding of what is necessary, such as telephone service; and allows for the impact of income taxes and tax credits. This workshop explores the need for the standard and how to use it in your own organization or life.

Contents
Workshop Agenda
Self-Sufficiency Standard Factoids Handout
Facilitator's Notes

Workshop Agenda: How Much Is Enough? The Self-Sufficiency Standard

Conversation (20 minutes)
The minimum wage: How'd we get here?

Exercise (20 minutes)
Does it add up? Does the minimum wage still make sense?
Calculate whether or not you think the family in the profile your group receives will make it, given the wages the family is earning and the assistance it is receiving.

Conversation (10 minutes)
What's the problem with the minimum wage? Make a collective list of expenses that the minimum wage and public assistance are not taking into account. What might a more just minimum wage and public assistance program take into account?

Conversation (20 minutes)
Myths of the minimum wage:
Myth 1: Working steadily at a job is enough to stay out of poverty.
Myth 2: Voodoo economics: the poor will get richer if the rich get richer, through the trickle-down effect.
Myth 3: Poor folks in the United States are better off than pretty much everyone else in the world.
Myth 4: Only the poor get public assistance.

Break for Ice Cream (15 minutes)
Play the Oprah Winfrey interview with Ben Cohen. (5 minutes)
Myth 5: Good wages are bad business.
What might be some of the benefits for *businesses* of paying good wages?

Exercise (15 minutes)
Calculating self-sufficiency: How would you define a self-sufficiency or equity wage? Return to the chart made earlier, showing what minimum wage doesn't take into account. What else would we have to take into account for a truly just minimum

wage, one based on meeting minimum needs? Let's figure it out together for Rensselaer County:

By playing . . . THE WAGE IS RIGHT game show!

Wrapping Up, Moving to Action (30 minutes)

Myth 6: We can't do anything about it.

1. There are alternatives. The Self-Sufficiency standard, which we just calculated, is one. But how else could we calculate minimum basic supports? What kinds of institutions and policies might help get us there?

2. There are successes. Successful living wage campaigns (the living-wage sit-in at Harvard University, for example). Play a clip from the *Occupation* video (Razsa and Velez, En Masse Films, 2002).

3. What can we do? What's going on in our community that we can get involved in? If there's nothing going on, how can we start something? Brainstorming.

Self-Sufficiency Standard Factoids

A true revolution of values will soon cause us to question the fairness and justice of many of our past and present policies. We are called to play the good Samaritan on life's roadside; but . . . One day the whole Jericho road must be transformed so that men and women will not be beaten and robbed as they make their journey through life. True compassion is more than flinging a coin to a beggar; it understands that an edifice which produces beggars needs restructuring.

There is nothing to prevent us from paying adequate wages to schoolteachers, social workers, and other servants of the public. . . . There is nothing but lack of social vision to prevent us from paying an adequate wage to every American citizen whether [s]he be a hospital worker, laundry worker, maid or day laborer. There is nothing except shortsightedness to prevent us from guaranteeing an annual minimum—and livable—income for every American family.

—Martin Luther King, Jr., "Where Do We Go From Here: Chaos or Community?" (1967).

Rising economic tides do not lift all boats.

Adjusted for inflation, the minimum wage in 2000 was lower than in 1950, despite a decade of record-breaking economic growth.

Poverty is a childhood disease.

One in six children overall, and one in three African American and Latino children, are growing up poor, even by the (inadequate) official poverty measure. There is a strong link between the percentage of full-time workers being paid low wages and high child poverty rates.

The gender wealth gap is wide.

The typical minimum-wage earner is an adult woman. While women make up just under half the total workforce, two out of three minimum-wage workers are women.

The racial wealth gap is wider.

White non-Hispanic families had a median net worth (household assets, including home equity, minus debt) of $94,900 in 1998, while nonwhite or Hispanic families had a net worth of $16,400—one-sixth that of whites.

Inequality is bad for your health.

The United States is the richest nation on earth, but it is the only major industrialized nation *not* to assure health care for all its citizens, whether through a public, private, or mixed system. The United States ranks thirty-second in child mortality for children under five years old, tying with Cuba and Cyprus and behind Canada, Australia, Israel, Singapore, South Korea, and all of Western Europe. Lack of health insurance is generally associated with a 25 percent higher risk of death. Uninsured women are nearly 50 percent more likely to die four to seven years following an initial diagnosis of breast cancer than insured women.

The minimum wage is a poverty wage.

The federal minimum wage was enacted in 1938 to put a firm floor under workers and their families, strengthen the economy by increasing consumer purchasing power, create new jobs, foster growth in lagging regions, and prevent a "race to the bottom," with employers moving to the cheapest possible labor state. In 1938, the minimum wage brought a family of three with one full-time worker above the federal poverty line. [Talk about how the poverty line was calculated based on food cost as the highest cost of a household. Today, a household's heaviest financial burden is housing.] The current minimum wage doesn't bring one worker with one child above the federal poverty line. The minimum wage is now a poverty wage.

The poverty line is inadequate.

In addition, the federal poverty line drastically underestimates people's actual needs. The Ms. Institute estimates a "minimum-needs" budget for a single worker without

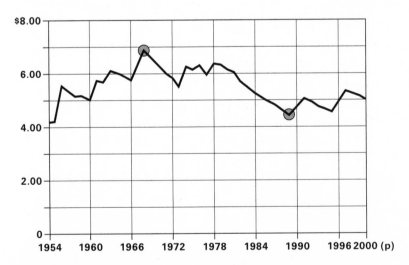

Figure C.1
The Federal Minimum Wage
The value of the minimum wage, 1954–2000 (in 1999 dollars).

employment health benefits at $16, 550 (or $7.96 an hour). For a single parent with two children (again without health benefits), the Ms. institute calculates a minimum-needs budget of $32,999 (or $15.86 an hour). The self-sufficiency wage varies by geographic area, but in Rensselaer County, the self-sufficiency wage for a single worker is $7.06 an hour ($1,129.60/month); it is $13.31 an hour ($2,129.60/month) for a single parent with an infant and $17.71 ($2,833.60/month) for a single parent with an infant and a preschooler.

Sources

Minimum-wage information and images adapted from the AFL-CIO report, *Raising the Minimum Wage: Talking Points and Background*, http://www.aflcio.org/mediacenter/resources/upload/minimumwagefactbooklet.pdf (accessed May 18, 2009). Other information adapted from *The State of Working America, 2000–2001*, http://www.epi.org/pages/books_swa2000_index2/ (accessed April 18, 2009). Self-sufficiency wages were calculated for 2002.

Facilitator's Notes

How Much Is Enough? The Self-Sufficiency Standard

Conversation Point (20 minutes)

The minimum wage: How'd we get here?

How much is the minimum wage? [$5.15]

Where did it come from?

The 1938 Fair Labor Standards Act (FLSA) included minimum-wage, overtime pay, and child labor provisions. In 1938, only 12 million workers were covered by the 25 cent minimum wage. Today, more than 113 million full- and part-time workers are covered.

Source

Department of Labor, Employment Standards Administration, "History of ESA," http://www.dol.gov/esa/about/history.htm.

What else was going on at the time?

• *Great Depression*: In 1933, one in four American workers were out of a job, and 20 percent of New York City schoolchildren were underweight and malnourished.
• *The New Deal*: Franklin D. Roosevelt's New Deal, a program of relief, recovery, and reform, lasted from 1933 to 1939 (Securities and Exchange Commission, Social Security, WPA, etc.).

How was it calculated?

The minimum wage is currently not indexed to any standard (e.g., cost of living, inflation, or poverty line). At its current level, the minimum wage will bring only a small number of families above the poverty line, if that family has one wage-earner working full-time. Many argue that the poverty line itself is outdated and needs to be updated.

Molly Orshansky developed the federal poverty line in 1963–1964. It is based on the "thrifty food plan," the cheapest of four food plans developed by the U.S. Department of Agriculture. Orshansky knew that families of three or more persons spent about one-third of their after-tax income on food. She then multiplied the cost of the USDA economy food plan by three to arrive at the minimal yearly income a family would need. Using 1963 as a base year, she calculated that a family of four consisting of two adults and two children would spend $1,033 for food per year. Using her formula based on the 1955 survey, she arrived at $3,100 a year ($1,033 x 3) as the poverty threshold for a family of four in 1963.

Source

Jessie Willis, *How We Measure Poverty: A History and Brief Overview*, http://www.ocpp .org/poverty/how.htm (2000).

Talk about differences in expenses between then and now. Is food a family's biggest expense? If not, what is?

Why did they do it?

The federal minimum wage was enacted to put a firm floor under workers and their families, strengthen the economy by increasing consumer purchasing power, create new jobs, foster economic growth in lagging regions, and prevent a "race to the bottom," with employers moving to the cheapest possible labor state.

Exercise (20 minutes)

Does it add up? Does the minimum wage still make sense?

Break the group into six small groups (using money). Give each group one of the family profiles and ask them to calculate whether or not they think the family will make it, given the wages the family is earning and the assistance it is are receiving. Be sure to look for holes in the profiles: What did we overlook? Where are the potential catastrophes? Is there potential help we didn't see?

Conversation Point (10 minutes)

The smaller groups reform as one big group. What's the problem? What's the problem with the minimum wage? Make a collective list of expenses that the minimum wage and public assistance are not taking into account. What might a more just minimum wage and public assistance program take into account?

Conversation Point (20 minutes)

Myths of the minimum wage. (Skip if short on time, have packets to hand out, or move until after ice cream if people are getting antsy.)

Myth 1: Working steadily at a job is enough to stay out of poverty.

In 1938, the federal minimum wage brought a family of three with one full-time worker above the poverty line. It was calculated by. . . . But now, a household's heaviest financial burden is housing. . . . The minimum wage doesn't bring one worker with one child above that the poverty line line. The minimum wage is now a poverty wage.

Myth 2: Voodoo economics: the poor will get richer if the rich get richer, through the trickle-down effect.
Actually, the United States has one of the most extreme inequalities in wealth of the industrialized countries, and it's getting worse, not better.

Inequality is bad for your health.

The United States is the richest nation on earth, but it is the only major industrialized nation *not* to assure health care for all its citizens, whether through a public, private, or mixed system. The United States ranks thirty-second in child mortality in children under five years old—we're tied with Cuba and Cyprus, and behind Canada, Australia, Israel, Singapore, South Korea, and all of Western Europe. Lack of health insurance is generally associated with a 25 percent higher risk of death. Uninsured women are nearly 50 percent more likely to die four to seven years following an initial diagnosis of breast cancer than insured women.

The gender wealth gap is wide.

The typical minimum-wage earner is an adult woman. While women make up just under half the total workforce, two out of three minimum-wage workers are women.

The racial wealth gap is wider.

White non-Hispanic families had a median net worth (household assets, including home equity, minus debt) of $94,900 in 1998, while nonwhite or Hispanic families had a net worth of $16,400—one-sixth that of whites.

Myth 3: Poor folks in the United States are better off than pretty much everyone else in the world.
Other countries do a much better job of eliminating poverty because their redistributive policies are better.

Poverty is a childhood disease.

One in six children overall, and one in three African American and Latino children, grows up poor—even by the (inadequate) official poverty measure. There is a strong link between the percentage of full-time workers being paid low wages and high child poverty rates. A recent Children First report calculated that the United States is second among industrialized nations in childhood poverty rates (behind the Russian Federation), with 39 percent of American children living at or near the poverty line.

Myth 4: Only the poor get public assistance.
Another reason the United States doesn't do a good job eliminating poverty is because we subsidize the wealthy—and if you include subsidies to corporations (i.e., bank bailouts, business incentives, etc.), the situation gets even more disconcerting.

Break for ice cream (15 minutes)
We're going to use this ice cream break to continue the conversation on the myths in circulation about minimum wage versus fair or livable wages.

Myth 5: Good wages are bad business. (Responsible businesses often flourish.)
A real-world example:
Until recently, Ben & Jerry's had a salary cap that limited the top executive's salary to seven times that of the lowest-paid employee, the $8 an hour scooper (though this was a little bit PR—the story of the salary cap masks the owner's wealth in stock shares). They also buy their dairy products locally—at or above the market rate—in an effort to help Vermont family farms and to provide dairy products free of rBGH, a cow growth hormone. The company gives away 7.5 percent of its pre-tax profits to charity, the highest of any American company (the average is 1 percent). Some of this is shrewd business, not just good intentions. For example, their philanthropy and unusual business model resulted in millions of dollars worth of free advertising and strong product loyalty.

What might be some of the benefits for businesses of paying good wages?

Higher wages and good benefits help companies retain employees, leading to reduced turnover and absenteeism.
Many responsible businesses report improvements in the quality of products and services.
With higher wages, workers have more purchasing power, which buoys the economy. Ben Cohen has gone on to found Sweat X, which makes "sweatshop-free clothes for large organizations like universities." Play Oprah's tape. (5 minutes)
(How does globalism throw a new wrinkle in the whole thing? Connect back to original aim of the 1938 *federal* wage being to prevent a race to the bottom. How might an international minimum wage be calculated?)

Exercise (15 minutes)
Calculating self-sufficiency   How would you define a self-sufficiency or equity wage? Return to the chart developed earlier showing what the minimum wage doesn't take into account. What else would we have to take into account for a truly just minimum wage, one based on minimum needs? Let's figure it out together for Rensselaer County:

THE WAGE IS RIGHT game show.

Last myth: We can't do anything about it. (30 minutes)
1. There are alternatives: the self-sufficiency standard, which we just calculated, is one. Other solutions include developing a minimum wage based on minimum needs (through living-wage campaigns) and working toward a guaranteed minimum income in the United States. Also important would be universal health care, paid caregiving, affordable housing, unions and labor laws, pay equity, education and training, savings and investment.

2. There are successes. Mention successful living-wage campaigns (use the *Occupation* video, or a clip of it from Oprah).

3. What can we do? What's going on in our community that we can get involved in? If there's nothing going on, how can we start something? Brainstorming.

Beat the System: Surviving Welfare

Workshop presented May 3, 2003, as part of the Pavilion Skillshare, Forum, and Picnic, Troy, New York (Andrew Lynn and Anne Marie Lanesey, organizers)

Authors: Women at the YWCA Making Social Movement

Description

Women at the YWCA Making Social Movement (WYMSM) was designing Beat the System: Surviving Welfare, an educational board game about social and economic justice issues in New York State, when we were invited by activists Andrew Lynn and Anne Marie Lanesey to participate in Pavilion.

At the Pavilion event, WYMSM presented progress on designing the game, which seeks to empower its players by providing insight into, and survival skills for dealing with, the human and social service systems. WYMSM involved Pavilion participants in the unique research process used by the group to evoke shared experiences and life stories for the game by designing the following popular education exercise.

Contents

Workshop Agenda
Summary Points: Welfare "Reform"
Skit Preparation Instructions
Composites
Discussion Questions
On the Composites: WYMSM's Research

Workshop Agenda

- Introduction to WYMSM (5 minutes)
- Small group skit preparation (20 minutes)
- Skit performances (20 minutes)
- Large group discussion (15 minutes)

Summary Points: Welfare "Reform"

• A key premise of the welfare reforms of 1996, which ended the Aid to Families with Dependent Children (AFDC) program, was that state governments, if given appropriate discretion, are best positioned to ensure the well-being of low-income families with children. This is reflected in the design of Temporary Aid to Needy Families (TANF), which allows states broad flexibility in the expenditure of federal and state welfare funds.

• One of the harshest aspects of the TANF program is the lifetime limit (60 months) to the receipt of federal welfare benefits.

• Many families with children who are eligible for help from TANF are not getting it. In the first two years after welfare reform (during an economic boom), the number of people eligible for TANF dropped only 5 percent, but participation in TANF dropped 23 percent.

• Despite all the rhetoric about marriage, sixteen states and the District of Columbia retain policies from AFDC that discriminate against two-parent families and effectively exclude them from the welfare safety net.

• Despite all the rhetoric about work, welfare rules in most states create an incentive for parents to leave their jobs in order to access food assistance, subsidized housing, and health care for their families.

• Large numbers of immigrants are effectively excluded from the welfare safety net.

Sources

"Kicked Off, Kept Off: How TANF Keeps Low-Income People Poor" (Washington, DC: The National Campaign for Jobs and Income Support, February 14, 2002).

"States Behaving Badly: America's 10 Worst Welfare States" (Washington, DC: The National Campaign for Jobs and Income Support, February 22, 2002).

Skit Preparation Instructions

1. Read your composite story, and note down the main points and whatever themes you see.

2. Discuss the small group discussion questions to identify one point or theme to illustrate in your skit.

3. Develop a plot and characters for your group based on that point or theme.

4. Decide who will take on which character and how the action will unfold.

5. Gather any props you might need.

6. Rehearse once!

7. Don't worry, we love improv! But remember that each group will have only 3–5 minutes to present!

Composites

• *Sandra* is a thirty-four-year-old mother of three who is reapplying for assistance just after TANF's five-year lifetime limit was put into place. Although she had been briefly on public assistance prior to 1996, she finds herself entrapped by the bureaucratic requirements this time around. First she had to travel to another city to get her first child's birth certificate within a ten-day time period. Second, when she asked her landlord for documentation of her rent receipts, she learned that he did not take Department of Social Services (DSS) applicants, so she realized she would have to find another place to live. Because of lead exposure restrictions, she had to find a place built after 1976, which she had difficulty doing, given the housing stock in her area. Third, she had to produce the names of her children's fathers. Finally, she had to document proof of income, but she had to leave her job because she had left so often to do errands related to her attempt to access public assistance. She ends up in a homeless shelter with her children.

• *Maria*, a forty-two-year-old recent immigrant, received her medical degree in her home country of the Philippines. She has taken the most available job she can find, as a live-in nanny in Westchester County, hoping to find time to study for her U.S. medical boards. She sends a large portion of her salary back to the Philippines to support her three children, who remain there under the care of their grandmother. She is afraid she won't be able to afford to redo her internship and residency after she passes her exam.

• *Harry* is a forty-five-year-old factory worker who was laid off when his employer moved offshore. When a relative refers him to unemployment, he is denied for not filling out the paper work correctly because he is functionally illiterate. He doesn't understand why he was denied unemployment compensation after working many years for the same employer. The denial comes in the form of a lengthy, legalistic letter that he gets a friend to read. Understanding that his options with unemployment are exhausted, he then tries to apply for assistance through DSS. Having no previous experience with social services, he is unaware of programs for which he might be eligible, such as emergency food stamps. No one informs him about his eligibility because they figure he will be able to find the information himself or will just go out and get another job. Nor does anyone figure out that he is not literate. He is again denied benefits, thus falling between the cracks of the two systems. Out of pride, fatigue, and frustration, he does not return to either system but instead becomes homeless. He does not seek a homeless shelter, where he might have obtained advocacy and support, but instead lives on the streets. This causes him to fall into a depression, and he begins to show symptoms of mental illness.

• *Karen* is a fifty-two-year-old woman who has struggled with mental health issues (particularly depression and bipolar disorder) for twenty tears. Though she has suffered several nervous breakdowns, she managed to maintain state employment as a keyboard operator and in data entry and forms processing for much of the past thirty years. Recently, though, she was institutionalized for nine months, and is trying to reenter the workforce. She finds an advocate from outpatient referrals and goes to the DSS to apply for unemployment, Medicaid, and food stamps. She becomes increasingly frustrated with DSS after a five-hour wait to see a caseworker, the depersonalized treatment she faces when she finally gets to see someone, and her general feelings of dependency and helplessness. The advocate does her best to help her navigate the system, but Karen is certain that she is being discriminated against because of her mental illness.

• *Rebecca*, a nineteen-year-old mother of one child, dropped out of high school in tenth grade during her pregnancy. She has been living with her parents, and she is thus ineligible for aid until age twenty-one unless she is legally emancipated, her parents go onto public assistance, or they pass away. However, her father is abusive toward her, and her parents become unable or unwilling to support her and her child. At this point in her life, Rebecca wants to complete her GED and seek higher education in order to better support herself and her son. Because she can document the abuse, she is put in a temporary group home, where a sympathetic caseworker refers her to Community Maternity Services in Albany, informing her that this agency is one of her best options because it comprehensively serves people in her position. In this highly structured program, she finds not only housing but child care, job training, and life skills classes, and a GED class.

• *Andrew* is a thirty-three-year-old intravenous drug user recovering in a MICA program. He has hepatitis C. Although he receives Medicaid, the federal health insurance program for indigent people, his bills exceed what Medicaid pays for the expensive pharmaceutical drugs that are essential to his liver functioning. His drug bills run about $4,000 annually. Seven months into the year, his Medicaid coverage runs out, and he is denied for the rest of the year. He has to switch to one primary care physician who works for a clinic that will accept just what Medicaid pays and not bill him for the balance, and drop all his other specialists.

• *Billy* is a twenty-two-year-old high school dropout who works full-time on the night shift at Burger King. His live-in girlfriend, Peggy, is four months' pregnant. He is hoping to become a welder, so he contacted the Capital District Educational Opportunities Commission (EOC) to enroll in a ten-week course, which should make him employable a month or two before his girlfriend gives birth. Unfortunately, when he arrives at his first class, he learns that he needs his GED before he can take the welding course, and is referred to another organization, where he is told it will take him from three to six months to complete GED training and testing. If he doesn't find a higher-paying job, he and his girlfriend will have to move into separate apartments so that she can apply for TANF to get the support she needs to raise a child.

• *Jane*, a twenty-two-year-old single mother, has one child, Johnny, who develops respiratory syncytial virus infection, goes to the emergency room, and is hospitalized. When he leaves the hospital, the doctor says he must stay home a couple of days. Jane works twenty hours a week at a new job at CVS. She doesn't have anyone she can depend on to help her by watching Johnny while she is at work. When she calls her new boss to explain her situation, she is told to come in or she will be fired. Jane feels she has to make a choice between her job and her child. She chooses to stay home with her child and is fired. She goes to Project Hire to tell her caseworker that she lost her job. The worker informs her they have to sanction her for ninety days. By law they cannot sanction her child, but she is left with no cash benefits and no support to seek more sustainable employment.

• *Daleesha* is sixteen years old and has an infant daughter. She lives with her forty-two-year-old father, who works a full-time second-shift state job in the capital and is able to cover the mortgage and include Daleesha in his health insurance coverage as a dependent (until she is eighteen). Daleesha's high school has recently started an experimental teen parenting program that provides day care so that she can stay in school, and though she sometimes struggles to stay focused with very little sleep and more responsibility than her friends, she is maintaining a 2.8 GPA. She hopes to apply to colleges in the fall, but she's heard recently that because of budget cuts put into place by the new Republican city council, the teen parenting program may be discontinued in her final year.

Discussion Questions

Small Group Discussion Questions

1. What options does your character have?
2. How well is the system supporting a sustainable life for your character and her or his family?
3. What decisions might your character make, and why?
4. What might the consequences of these decisions be?

Large Group Discussion Questions

1. How well is the system supporting a sustainable life for the people represented in the composites?
2. How well does this arrangement allow your characters to take care of themselves, their family, and earn an income?
3. What about independence and self-respect? Any trade-offs?
4. How can these characters organize to get a better deal? What governmental and social supports do they need?

On the Composites: WYMSM's Research

During the last several months, WYMSM has been using an interviewing and compositing method to gather and combine information from a wide variety of people about their experiences with the human and social service systems. The composites you read today are a product of that research, and have been distilled from weeks of interviewing and discussion by the WYMSM team.

These stories, based on people's real lived experiences, will be used for a variety of purposes, including workshops like today's. We will also use the composites to help us develop characters and scenarios for the board game we are creating called Beat the System: Surviving Welfare.

WYMSM is a collaborative community group that uses technology as a tool of social change.

Technology, Power, and Resistance

First offered at the National Network to End Domestic Violence Safety Net Conference: Training of Trainers v6, Pittsburg, Pennsylvania, Tuesday, August 12, 2008

Authors: Virginia Eubanks and Christine Nealon

Description

Safety Net, a technology safety project of the National Network to End Domestic Violence (NNEDV), holds an annual Training of Trainers conference. At the conference, the network discusses how technology is being used to victimize domestic violence survivors and also how it is being used positively in advocacy and organizing efforts. In 2008, I was asked to deliver a keynote talk about my work on technology, violence, and women's citizenship at the conference. They also asked that Christine Nealon and I develop a hands-on workshop dealing with some aspect of our popular technology work.

We created the "Technology, Power, and Resistance" workshop to help advocates and activists think through how technology impacts their day-to-day work and their broader social justice concerns. It was important to us to help participants think about the relationship between technology and power in a broad and structural way rather than seeing technology as simply a tool for their work or a threat to their clients.

Contents
Workshop Agenda
About Our Knowledge, Our Power (OKOP)
About the Facilitators
Power Inventory Handouts
Facilitator's Notes

Welcome to "Technology, Power, and Resistance"!

Workshop Agenda
8:45–8:50 a.m. **Welcome!**
Ground Rules
8:50–9 a.m. **Icebreaker—Ball Throw**
9–9:05 a.m. **Introduction**
9:05–9:25 a.m. **Technology and Social Justice Inventory**
✋ Hands-on issue mapping
9:25–9:40 a.m. **Identify Confrontable Issues**
(Discussion)
9:40–9:45 a.m. **The "Big 5"**
(Break into content groups)
9:45–10:10 a.m. **Power Inventory**
10:10–10:15 a.m. **Wrap-up, Evaluation**

About Our Knowledge, Our Power
Our Knowledge, Our Power (OKOP) draws on values of respect, local expertise, grassroots process, and true democracy to make real, meaningful change in the terribly unfair and exploitative system of public assistance in New York State, which keeps people dependent and poor.

We try to counteract the alienation and lack of respect that public assistance beneficiaries often encounter by sharing information and resources to bridge the gap between public assistance recipients and workers, politicians, and the general public.

We believe we can bridge this gap by drawing on our collective power to:

- Empower people most directly impacted by the social service system;
- Enlighten people about their rights in the system;
- Provide tools for navigating the system;
- Help with the application process;
- Facilitate knowledge sharing; and
- Share our needs, challenges and aspirations with other members of the community.

OKOP runs entirely on donations, small grants and volunteer labor. If you can donate time, money, or supplies to help people struggling to meet their basic needs claim their economic human rights, please contact Virginia at Virginia@populartechnology. org or (518) 892–6697. For more information on our organization, see http://www .populartechnology.org.

About the Facilitators

Virginia Eubanks joined the Department of Women's Studies at the University at Albany, SUNY in 2004 after completing her Ph.D. in Science and Technology Studies at Rensselaer Polytechnic Institute. Eubanks came to her research on technology, women's poverty, and citizenship in the United States through a history of activism in community media, technology center, and antipoverty movements. She is currently working on a book entitled *Technologies of Citizenship: Women, Inequality, and the Information Age*. Eubanks also co-founded the Popular Technology Workshops, which serve as a place for people struggling to meet their basic needs to come together in order to define their problems and develop their own solutions. The workshops are grounded in the idea that people closest to problems have the most information about them and are most invested in creating smart solutions. More information about her is available at http://www.populartechnology.org/Virginia/.
Christine Nealon calls on her more than fifteen years of experience with various nonprofits, including the YWCA, Catholic Charities, and Equinox, to fuel her passion for popular education and grassroots organizing. She continues to blend the academic world and organizing for human rights in the local, national, and global arenas while teaching the WORLD courses at Russell Sage College in Troy, New York, engaging in research with Virginia Eubanks, Ph.D., and completing her graduate work at the State University of New York at Albany in the Women's Studies Department.

Power Inventory

This three-part power inventory is used to take stock of the kinds of power available to your group when undertaking a technology and social justice project. The inventory asks us to consider:

Does your group have some forms of power?
Where and in what fields do you have the most power?
How can you use that power to achieve social and economic justice:

⇨ Within the organization?
⇨ Within the specific field/issue?
⇨ In our communities?

Part I: Name the Change You Want to Make
We want to

_____

What's the relationship between your technological issue and the work you (and your organization) already do?
How does this technological issue impact the people you serve? Directly? Indirectly? What are the barriers to successfully confronting this issue? Who benefits from this issue existing?

Part II: Identify resources you have to make that change

	Financial resources: Money, property, capital	Power: Influence, elected office, votes, community base	People: Personnel, stakeholders, employees	Skills: Organizing, new ideas and actions, bilingual, typing	Other
Organizations					
Businesses					
Your job					
Friends and family					
Community/ neighborhood					
Politics					
Other					

Source: Adapted from Hope and Timmel, *Training for Transformation* (ITG Publishing, 1999).

Part III: Identify Your Limitations, Develop Strategies to Fill Gaps
1. Discuss: Where does your funding come from? Are there ways that your funding source impacts your goals or methods? How much time do you spend responding to the needs of your funders rather than the needs of the people you serve?
⇨ Write STRATEGIES for strengthening ORGANIZATIONAL SELF-DETERMINATION
2. Discuss: How well do you work with existing community leaders? Does your organization help foster leadership among the people it serves? Are there ways that your organization may be diminishing the growth of grassroots organizing/leadership?

⇨ Write STRATEGIES for strengthening INCLUSIVITY and SHARING POWER

3. Discuss: In what ways are the staff of your organization separated from the people you serve because of the following: the status and pay of staff; the professionalization of the work; the role of your organization in the community; the interdependence of your work with governmental agencies, businesses, foundations, or other nonprofit organizations?

⇨ Write STRATEGIES for strengthening DEMOCRATIC PROCESS

4. Discuss: In what ways have your ties to governmental and community agencies separated you from the people you serve? In what ways have those ties limited your ability to be "contentious"—to challenge the roots of violence and undemocratic and abusive practices?

⇨ Write STRATEGIES for strengthening COMMITMENT TO SOCIAL CHANGE

Source

Adapted from INCITE! Women of Color Against Violence, *The Revolution Will Not Be Funded* (Cambridge, MA: South End Press, 2006).

Facilitator's Notes

Icebreaker—Ball Throw (10 minutes)

1. What technology can you (personally) not live without?
2. If you had the power to remove one technology from the world, what would it be?
3. What technologies can your organization not live without?
4. What technology weakens or threatens your organization?
5. Four goofy questions

Introduction to Workshop/History of Work (5 minutes)

This workshop affords an opportunity to think broadly about technology and social justice work but does not answer specific questions about technology per se, because that is covered well in the rest of the conference. Our role as one of the first workshops at the conference is to ask big questions about the relationship between technology and social, political, and economic justice and human rights.

Structure: Heart Head Feet (Start with your own experience; Analyze it collectively; Identify areas for action)

Technology and Issue Inventory (20 minutes)

In the technology and issue inventory, ask participants to choose from our pile, or create on blank paper the following: (1) a pictures of a technology that is interesting or important in regard to the work you care about (e.g., computer, credit card, medicine) and (2) a picture or title that describes an issue you feel passionately about (e.g., violence against women, the environment, voting). These two things should somehow be related.

Then, everyone sticks her two pictures on a large piece of butcher paper along one of the walls of the room. When you have finished putting your contribution up, use markers to draw connections and relationships between different technologies and issues, or write descriptions and thoughts that come to mind when you look at the "mural" the group is creating.

Discussion: Identify Issues You Want to Work On (15 minutes)

1. What picture has emerged?
2. Surprises?
3. What are the "centers of gravity"?

Identify the "Big 5" Issues (5 minutes)

List "Centers of Gravity" on a flip chart, and give every participant five sticky dots to vote with. They can give five different issues one dot each, or one issue they feel particularly strong about all five of their votes. The top five issues become our "Big 5" issues, and we'll break into five groups to discuss them.

Once you are in a content group, get practical about specific problems you want to solve (for the purposes of today's workshop).

Three qualifications: You must be passionate about the problem; the problem must be solvable, and the problem must concern social, cultural, political, or economic justice.

Power Inventory (30 minutes)

Hand out and describe the power inventory. If you run out of time, give groups parts 2 and 3 to take home to their organizations, and just complete part 1 during the workshop.

Questions we are answering:

Does your group have some forms of power?
Where and in what fields do you have the most power? How can you use that power to achieve social and economic justice?

Practical Goals:
Identify changes you want to make.
Identify resources you currently have to make change.
Identify your limitations and develop strategies to fill in gaps.
(Collect contact info while participants are working.)

Wrap-up, Evaluation (5 minutes)

There is not enough time to discuss all of the great material that's been brought forward here. So, we're going to ask that you make one commitment toward creating the change you want to see in the world. Or write on the postcards we're distributing, what was one thing that you got from today that you want to hang on to? We will send you the postcard in six weeks to remind you of the commitment you made here today.

General Notes on Creating a Popular Technology Workshop

Prepared for community workshop/class instructors at the YWCA of Troy-Cohoes by Virginia Eubanks, Spring 2002

We generally are looking for three orientations or methods in popular technology workshops:

1. *Popular Education* Focus in popular education is on education as empowerment. It combines grassroots research projects with adult education and social/political action in order to provide more equitable control over the means of both intellectual production (i.e., research for the people, by the people) and the possibility of social change. Workshops are intended to provide a space for nurturing collaboration, growing critical collective self-consciousness, and mobilizing people to take action in their own lives and community.

2. *"Dig Where You Stand" History* Closely connected to popular education, "Dig where you stand" is an approach that uncovers and represents what ordinary people felt, thought, and tried to accomplish, often discovering or highlighting parts of a community's past that have been hidden in the traditional rendering of history. "Dig where you stand" challenges the idea that history is made by elites—in terms of who history is about (i.e., working people, social activists, or individuals, as opposed to presidents, armies, and ruling classes), who collects and interprets it (i.e., labor unions and ordinary folks versus academic institutions), and whom it should be for. It also challenges where "history" takes place. In the words of Mary Harris "Mother" Jones, "Dig where you stand and you will find it."

3. *Constructivist/Activist Learning* Finally, we believe that, though there are many different kinds of learning styles, many people learn best by doing, through hands-on, concrete projects that have direct relevance to their daily lives. Most workshops begin with contextualization of/on the subject (i.e., storytelling, slide shows, short lectures), then cycle through other means of getting at the same information. So, for example, the second workshop in the series began with a slide show about the different kinds of employment women in the United States have historically held. Then the group cycled through a number of mapping activities intended to get at the same information in a more personal, concrete ways: drawing family migration maps (historical context) and time studies (personal context). Finally, we think it is important to leave sufficient time at the end of all workshops for reflection—often accomplished by "walking through" the visuals and content that was created during the workshop. So, generally, the YWCA needs 20 minutes (5–10 minutes at the beginning and 10–15 minutes at the end) to do introductions and evaluations, and we generally suggest no more than 30–45 minutes in lecture or storytelling mode (can be broken up), which would leave you with 45 minutes to an hour of hands-on (or feet-on, in the local history sense) activity time.

Women's Economic Empowerment Series Schedule (2002)

The Sally Catlin Resource Center Presents:

The 2002 Women's Economic Empowerment Series
At the YWCA of Troy-Cohoes

• Wednesday, July 10, 2002, 4–6 p.m.

The Richness of Our Roots: Family Portraits and Genealogy

Reclaiming our families' economic histories

In conjunction with the Rensselaer County Historical Society (RCHS)

• Wednesday, July 17, 2002

What's Women's Work?

Mapping labor and power from the past to the present

In conjunction with the Community Outreach Partnership Center (COPC) at RPI

• Wednesday, July 24, 2002, 4–6 p.m.

Story Swapping

Songwriting, storytelling, and spoken word for social change

In conjunction with singer-songwriter Jenrose Fitzgerald

• Wednesday, August 7, 2002, 4–6 p.m.

Rediscovering Troy's Shared History

Uncovering and telling our communities' stories

In conjunction with the Capital Region Underground Railroad History Project

• Wednesday, August 14, 2002

How Much Is Enough? The Self-Sufficiency Standard

Meeting basic needs, and beyond . . .

In conjunction with the Women's Center for Education and Career Advancement (WCECA)

• Wednesday, August 21, 2002, 4–6 p.m.

Who's Counting? Global Economic Development

Women's unpaid and underpaid work in the global economy

In conjunction with Women at the YWCA Making Social Movement (WYMSM)

• Wednesday, September 11, 2002, 4–6 p.m.

Banking on the Future: Women's Investment

Community banking, individual development accounts, and microloans

In conjunction with Affordable Housing Partnerships

• Wednesday, September 18, 2002, 4–6 p.m.

Building Resource Networks in the Community

Community-government partnerships for a rich future

In conjunction with Women at the YWCA Making Social Movement (WYMSM)

• Wednesday, October 2, 2002, 4–6 p.m.

Celebrating Social Justice

The tree of life and our common future

Women's Economic Empowerment Series Closing Party!

All events are followed by a creative arts workshop on the following Monday. Join us in the Sally Catlin Resource Center, Mondays at 6 p.m., to continue each Wednesday's discussion.

All events meet at the YWCA of Troy-Cohoes, 21 First St., Troy, NY 12180.

The Sally Catlin Resource Center: Where women craft lives they want for themselves.

The YWCA of Troy-Cohoes is proud to announce . . .
The 2002 Women's Economic Empowerment Series!

21 First Street
Troy, New York 12180
ywca-troy.org

274-7100
274-2572 (fax)
press@ywca-troy.org

EVENT

2002 Women's Economic
Empowerment Series

"The Richness of Our
Roots: Family Portraits
and Genealogy"

DATE

July 2, 2002

FOR
IMMEDIATE RELEASE

FOR
MORE INFORMATION

Christine Nealon
(518) 274-7100 ext #111

Virginia Eubanks
(518) 274-7100 ext #113

On Wednesday, July 10, the Sally Catlin Resource Center at the YWCA of Troy-Cohoes will present the first workshop in their 2002 Women's Economic Empowerment Series. In collaboration with the Rensselaer County Historical Society, the first workshop—entitled "The Richness of Our Roots: Family Portraits and Genealogy"—will explore our families' and community's economic history through exhibits, exercises, and shared discovery. This program coincides with two of the Historical Society's current exhibitions: "The Family Life Show" and "Climbing Your Family Tree," which celebrate families in Rensselaer County and investigate the steps involved in documenting your own family history. Join us at the Troy-Cohoes YWCA from 4–6 PM for this fascinating program.

As part of the YWCA's mission to empower women and girls and to eliminate racism, the women's economic empowerment series responds to American women's economic instability. For decades, women's work has been under-appreciated and often unpaid. Many working women balance family and multiple jobs just to meet their most basic needs. Minimum wage in the U.S. is so low that it is next to impossible to maintain self-sufficiency with just one job. Despite these grim facts, there is much ordinary people can do to increase women's economic empowerment and equality.

This July, the series will also include:

❑ *What's Women's Work: Mapping Labor and Power in the Past and to the Present,* Wednesday July 17th from 4 to 6 PM
❑ *Story Swapping: Songwriting, Storytelling and Spoken Word for Social Change,* Wednesday July 24th from 4 to 6 PM

All events are continued in a creative arts workshop at 6PM on the following Monday. All events meet at the YWCA of Troy-Cohoes, 21 First Street, Troy, NY. For more information, or to RSVP, call Christine Nealon at 518/274-7100, ext #111.

Suitable for all people, of all ages, from all walks of life, and **ABSOLUTELY FREE** to the public. Brought to you by the Sally Catlin Resource Center (*Where women craft the lives they want for themselves*), the Rensselaer County Historical Society, the Arts Center of the Capital Region, and the Troy-Rensselaer Community Outreach Partnership Center (COPC).

By increasing women's participation in economic decision-making, by strengthening their bargaining power in both the household and marketplace, and by increasing access to and control over economic and natural resources, we can increase social justice and economic vitality, empowering ourselves and other women in our community.

Contact the Popular Technology Workshops

If you are interested in hosting a popular technology workshop, or in obtaining copies of popular technology workshop materials, contact:

Popular Technology Workshops
P.O. Box 1613
Troy, NY 12181
Virginia@populartechnology.org
http://www.populartechnology.org

Appendix D: Popular Technology Projects Undertaken at the YWCA of Troy-Cohoes

The YWCA Community Technology Laboratory   An internal technology resource collaboratively designed, administered, maintained, and programmed by community members, YWCA residents, staff, and volunteers. The lab was the site of classes and workshops, and lab members produced an audio CD (with original music), Web sites, newsletters, a presentation on Harriet Tubman in Troy history for a local Underground Railroad conference, and an installation featuring oral histories called Sewing for Survival.

The YWCA Online Women's Resource Directory   A "woman-to-woman" technology mentorship program focused on researching and designing a Web-based database of community programs and resources in Albany, Schenectady, and Rensselaer counties.

Beat the System: Surviving Welfare   An educational software program/board game/popular education exercise based on original research by WYMSM. Beat the System raises social and economic justice issues to empower players by providing insight into the survival skills needed to navigate the human and social service system.

The Women's Economic Empowerment Series   A nine-part workshop series dealing with women's economic justice issues, including global economic development, the self-sufficiency standard, women's paid and unpaid work, local histories of protest, and songwriting for social change. Held over several months in the summer and fall of 2002, the series attracted significant community involvement, and became a base through which we recruited several new WYMSM members.

Community Asset Bank   A database-driven Web application intended to facilitate the identification and sharing of community and individual assets by connecting resourceful people in low-income neighborhoods. The community asset bank, prototyped by RPI information technology undergraduate Jes Constantine, draws on the asset-based community development model for sharing material and intellectual resources between local residents and institutions. Though the prototype was interesting, lack of resources, lack of time, and miscommunication between YWCA community members and members of the RPI community kept us from creating a final working version of the bank.

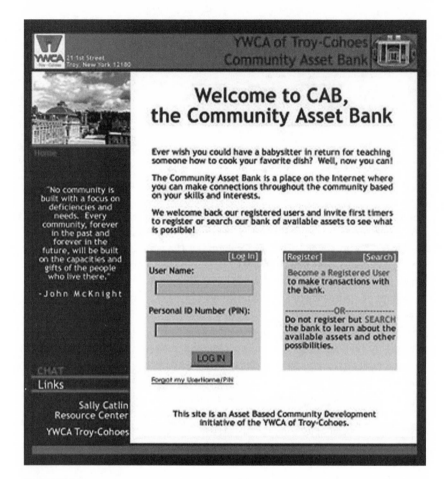

Figure D.1
Community asset bank, prototyped by Jessica Constantine.

Women at the YWCA Making Social Movement (WYMSM)   A diverse group of women from the YWCA community whose mission is to "use technology as a tool of social change" for community building, social networking, and resource sharing and development. Members met biweekly, acted as community peer educators at workshops and other public events, and designed a popular education exercise entitled Beat the System: Surviving Welfare.

Women, Simulation and Social Change Workshops   Two major workshops were held in the Sally Catlin Resource Center. The first was attended by twenty YWCA community members, covered the general idea of simulation and why it is of interest to community organizations and residents, and introduced participants to two major simulation games: The Sims and ROWEL's Life in the State of Poverty. The second workshop was attended by nine community members, who completed a popular education exercise called "The Tree of Life" in order to begin the process of storyboarding the simulation and to identify issues of concern to the community.

ROWEL's Exploring the State of Poverty Welfare Simulation   In honor of Hunger Awareness Day, in April 2002 WYMSM and other women from the YWCA organized and participated in a role-play simulation. More than ninety community members playing low-income family members were guided through a one-hour "month" of navigating the human and social service system. Participants then discussed the issues that emerged with YWCA residents.

Notes

Introduction

1. In the United States, we do not have terribly good language for talking about class, perhaps because we tend to believe that we live in a classless society. I have struggled in the book to find precise language that highlights the very real material and political impacts of class differences while respecting my collaborators' wishes and self-identification. I mostly use the term "poor and working-class" to describe women in the YWCA community. Though some members of the community would bristle at being described as "poor," as they see themselves as resourceful women with a rich array of skills and powers, many other members of the community took on the label "poor" or "working-class" as a political identity and marker of solidarity. In my experience, all poor people work, either for wages or as unpaid careworkers. Thus, I find "working poor" redundant, and do not use the term. While many scholars and policymakers use the phrase "low-income," to avoid the stigmatized connotations of "poor," I find it does not adequately describe the struggles that many poor and working-class people face—which stem only partially from the level of their income. Two sources that have clarified my thinking about the language we use in the United States to describe exploited classes of workers are Betsy Leondar-Wright's wonderfully insightful *Class Matters: Cross-Class Alliance Building for Middle-Class Activists* (New Society Publishers, 2005) and the work of the Poor People's Economic Human Rights Campaign (<http://www.economichumanrights.org>), which defines poverty as the inability to fully realize your economic human rights, such as food, housing, health care, education, communication, and a living-wage job.

Chapter 1

1. See, for example, "Getting Started: Roots of Radical Traditions," chapter 1 in Becky Thompson's lovely book, *A Promise and a Way of Life: White Antiracist Activism* (2001).

2. This is a practical problem with redistributive technology schemes faced by most nonprofits. Because it is assumed that access to a computer—any computer—is an

unmitigated social good, people donate old, barely functioning computers by the truckload to their local nonprofits. Ill-prepared and overworked staff then must find ways to repair or dispose of them, which can take time, energy, and other resources away from the organization's more mission-based programming. If staff and volunteers can in fact find time to refurbish the donations, it at best leaves the organization with low-functioning technology.

3. I use the term "critical ambivalence" here in Andrew Feenberg's sense—it is a part of an emergent critical theory of technology that popular technology education is intended to unleash. "Critical theory," Feenberg writes, "argues that technology is not a thing in the ordinary sense of the term, but an 'ambivalent' process of development suspended between different possibilities. This 'ambivalence' of technology is distinguished from neutrality by the role it attributes to social values in the design, and not merely the use, of technical systems. On this view, technology is not a destiny but a scene of struggle" (Feenberg 1990, 14). For Feenberg, critical theories of technology can be realized by rewriting "technical code" to inspire action that changes society and technology for the better, and to see the relationship between technology and people as intrinsic to modern social justice goals. The goal of popular technology is to undermine the myth of technological neutrality by finding "tipping points" that can guide development and design out of its "suspension between possibilities" and toward social justice. Like Paulo Freire, Feenberg sees critical theory as the rejection of value neutrality in analysis and a move toward more coherent and engaged understandings of the world.

4. This experience tends to support the claims of Freire, Horton, and other popular educators who insist that radical, "problem-posing" education truly starts when facilitators reflect back to participants the *contradictions* that shape their lives for analysis and action. I argue, therefore, that my collaborators' critical ambivalence in the face of technology is a sign of incipient analysis rather than apathy, fear, or ignorance.

5. The Reform Organization of Welfare (ROWEL) is a Missouri membership organization of low-income people and their allies dedicated to reducing poverty and prejudice. Since 1972, the organization has worked to improve the lives of Missouri's most vulnerable citizens through public education, community and legislative organizing, leadership development, advocacy training, political education, and direct-action campaigns.

6. The academic members of WYMSM were introduced to the YWCA through two students, Jes Constantine and Chitsunge Mapondera, who were involved in a very successful group-based internship project in the Sally Catlin Resource Center at the YWCA under the supervision of its director, Christine Nealon. Nancy D. Campbell, Ron Eglash, Nealon, and I then planned and held a series of four open meetings for YWCA community members under the name Women, Simulation and Social Change. These workshops explored the general concept of simulation, particularly

as it relates to policy forecasting. Over the four meetings, nearly eighty participants explored different kinds of simulations. When we asked participants how they would like to engage the rest of the community in dialog about their experiences surviving—and thriving—despite poverty, they chose the ROWEL format for a public event.

7. Although we paid a nominal stipend for women in the YWCA community to participate as "resource people," many made significant sacrifices to be there—taking the day off work or leaving early, traveling to the YWCA and juggling child-care commitments to attend. A staff member at the YWCA even managed to acquire a daylong furlough from the county jail for a woman who had participated in an early workshop so that she could play the role of landlord.

8. In retrospect, organizing this event foreshadowed the promises and perils of building social movements across class, race, and gender lines that we would struggle with throughout our time as WYMSM. One of the organizational representatives, the only man speaking at the press conference, went on for so long about public policy changes necessary to end poverty and support women and families on welfare that speakers from WYMSM—women who had actually experienced poverty and tried to survive on welfare—felt marginalized, excluded, and silenced. The newspaper coverage only showed pictures of white women, though the event had been extraordinarily diverse. The statewide antipoverty organization that had drafted the original press release for the event had neglected to mention that WYMSM existed at all, and had referred to our team of experts and activists drawn from the YWCA community only as women in need of "appropriate community resources," "independence [and] self-reliance." WYMSM spent a long meeting rewriting the press release to focus on community-building, interconnectedness, and grassroots expertise.

The original press release read:

The Statewide Emergency Network for Social and Economic Security (SENSES), Hunger Action Network of New York (HANNYS), Troy-Cohoes YWCA, and RPI Community Outreach Partnership Center will present Rowel's One Month in the State of Poverty Simulation starting with a press conference and ending with a debriefing and community meal from 3 to 6 PM at the YWCA, 21 First St., Troy.

This simulation is designed to help participants begin to understand what it might be like to be part of a typical low-income family trying to survive from month to month. The simulation will be run by women from the YWCA's Sally Catlin Resource Center, which promotes women's independence by encouraging women to develop greater self-reliance and linking them to appropriate community resources. This simulation kicks off a series of events at the YWCA dedicated to Women's Economic Empowerment.

After WYMSM's corrections, the new press release read:

The Statewide Emergency Network for Social and Economic Security (SENSES), Hunger Action Network of New York (HANNYS), Troy-Cohoes YWCA, Women at the YWCA Making Social Movement (WYMSM), and RPI Community Outreach Partnership Center (COPC) are partnering to present ROWEL's "Life in the State of Poverty" simulation, starting with a press conference and ending with an open dialogue and community meal, from 3 PM to 7 PM at the YWCA, 21 First St., Troy.

The ROWEL simulation is designed to help participants understand what it takes for a typical low-income family to survive from month to month. The simulation experience will be hosted and run by WYMSM—a community-building collaboration that uses technology tools and social networks to help women build awareness of local assets, share existing resources, develop new ones, and provide support to learn from each other's experiences—and other members of the YWCA community. This simulation experience will kick off a series of events by the Sally Catlin Resource Center dedicated to women's economic empowerment.

Nevertheless, we had more than ninety attendees, garnered press coverage and an editorial the next day in the local newspaper, fostered significant community dialog, and communicated an antipoverty and welfare rights message to an often apathetic middle-class public. It was the more personal, less visible triumphs that were most poignant for me. Three members of WYMSM, none of whom had ever spoken in public before, spoke out about their experiences battling poverty and the social service system to their peers, community members, and the press. YWCA residents, often seen as needy or deficient by the wider community, flourished in their roles as leaders and educators, reaching across the boundaries of race, gender, and class to make real connections based on mutual respect and lived expertise.

9. Lee discussed the idea of a "magical, mystical Negro" archetype in a talk at Yale University on February 21, 2002. See Gonzalez 2001.

10. From a speech on April 4, 1967, in Riverside Church, New York City. Quoted in Oates 1994 (436).

11. The good news is it is easier to be brave when you have company, something Barbara Smith calls "courage in concert" (personal communication, February 27, 2009).

Chapter 2

1. Benhabib 1996; Fraser 1989, 1997; Young 1990.

2. We are oppressed not as individuals but as members of social groups.

3. For more on technological due process, see Citron 2008.

4. Space prevents me from sketching even the broadest possible outline of the field of feminist science and technology studies. Luckily, there are fine review articles that do just that. The most current and comprehensive is Banu Subramaniam's 2009 article, "Moored Metamorphoses: A Retrospective Essay on Feminist Science Studies."

5. See, for example, Kolko 2000; Nakamura 2002, 2007; and Wright 2005.

6. The figure of the cyborg was probably best popularized by Donna Haraway in her generative text *Simians, Cyborgs and Women: The Reinvention of Nature* (1991).

7. See, for example, Cohoon and Aspray 2008; Cooks and Isgro 2005; Denner et al. 2005; Fuller and Meiners 2005; Jessup and Sumner 2005; Kekelis et al. 2005;

Margolis 2008; Margolis and Fisher 2002; Rosser 2005; Scott-Dixon 2004; and Tillberg and Cohoon 2005.

8. Hobson 2007; Nakamura 2007; Nelson,Tu, and Hines 2001.

9. Roberts 1998; Silliman et al. 2004.

10. See especially the chapter "At Your Service: Latinas in the Global Information Network," in Fusco 2002. See also Hossfeld 2001; Kumar 2001; Ng and Mitter 2005; and Wyatt et al. 2000.

11. Leslie McCall's meticulous empirical work on the information economy shows quite clearly that high and rising levels of inequality *among women* by race and class are a key characteristic of high-technology regions. Thus, regional development schemes that focus on high-tech industries trade commitments to broad-based equity for economic growth (McCall 2001, 190).

12. Information technologies, rather than being stable, neutral tools, are *technologies of citizenship*. For Barbara Cruikshank, technologies of citizenship are those discourses, programs and tactics "aimed at making individuals politically active and capable of self-government" (Cruikshank 1999, 2).

13. For example, Braman 2007; Lessig 1999; and Mayer-Schönberger and Lazer 2007.

14. For Paulo Freire, becoming critical is a movement toward more coherent understandings of the world. For example, in his posthumously published work *Pedagogy of Freedom: Ethics, Democracy, and Civic Courage,* Freire illustrates the failure of technical knowledge to create progressive social change by describing reading specialists whose technical advances in child literacy techniques have failed to reach the millions of children who remain illiterate by virtue of their race, ethnicity, gender, nationality, or class. Technology training curricula available in formal learning environments—schools, colleges, vocational training programs, and even many community organizations—often favor disarticulated skill training over generalized, problem-posing education (Macedo in Freire 1998, xii–xiv).

15. In this I follow Nancy Fraser's lead in exploring "actually existing democracy" in her 1997 work, *Justice Interruptus.* She writes in its introduction, "In the wake of 1989 [the collapse of the Berlin Wall] we have heard a great deal of ballyhoo about 'the triumph of liberal democracy' and even 'the end of history.' Yet there is still quite a lot to object to in our own 'actually existing democracy'" (69).

16. One of Donna Haraway's material-semiotic actors, the Modest_Witness can be seen as a stand-in for the kind of situated, provisional knowledge claim Franklin sanctions. In Haraway's words, the goal of the Modest_Witness is to "provid[e] good enough grounding" on which to build reliable, truthful accounts of important things in the real world "while eschewing the addictive narcotic of transcendental foundations" (Haraway 1997, 22). For more on this concept, and particularly, on the importance of modest knowledge claims to coalition building, see Bastian 2006.

17. See, for example, Haraway 1991, 1997; and Haraway and Goodeve 1999.

18. Popular technology focuses on developing forums in which all of us can become more critical technological citizens, in addition to teaching individuals technical skills and providing communities with technological resources. Though teaching technology skills is part of the approach, such skills are taught only when the need for them organically arises from broader social justice organizing goals. For example, video production might be taught to facilitate the production of a public service announcement, or word processing might be taught to produce handouts or leaflets.

19. Originally called the Highlander Folk School, Highlander was founded by Horton and Don West with the help of coal miners, farmers, and industrial laborers in East Tennessee.

20. There are exceptions, of course. One that has been particularly influential on my work is Dorothy Smith's "sociology for people" (Smith 1987, 1999, 2005).

Chapter 3

1. Some credit Donna Hoffman and Tom Novak with inventing the phrase. Novak and Hoffman (1998) and Compaine (2001) credit Morissett. Morissett himself expresses doubts about the term's origin in Compaine 2001. Nevertheless, the phrase reached popular currency after a 1998 NTIA report used it in its subtitle.

2. See, for example, Attali 2000; Brown 2001; Rischard 1996; and Yunus 2001.

3. The Clinton administration's attention to universal access to IT considerably anticipated the digital divide rhetoric, however. At the G7 Information Summit in early 1995, Vice President Al Gore explained that the Clinton administration's position on universal access was a commitment "to wire every classroom, every library, every hospital, and every clinic to the national and international information infrastructure." At the International Telecommunication Union World Telecommunication Development Conference later that year, he "called for all nations of the world to co-operate in building the Global Information Infrastructure founded on principles of universal access, the right to communicate, and diversity of expression" (Gore 1995 and Tarjanne 1995, quoted in Compaine 2001, 162–63).

4. National funding to bridge the digital divide peaked in 2001. That year the Technology Opportunities Program (TOP) received $42.5 million and the Community Technology Centers Program (CTC) received $65 million. In President Bush's fiscal year 2003 budget request, both programs were slated for elimination (OMB Watch 2002). In 2002, the CTC Program was funded at $32.5 million and TOP was funded at $15.5 million. Bush eliminated the program, but requested level funding of HUD's Neighborhood Networks program (at $20 million in FY 2002 and 2003). TOP was eliminated outright. In July 2002, the Senate restored the CTC and TOP programs

to FY 2002 levels in its FY 2003 Labor–Health and Human Services–Education Appropriations bill (National Institute for Literacy 2002). The Bush administration again called for both programs' elimination in the FY 2004 budget proposal (Komar 2003). The Obama administration's Broadband Technology Opportunities Program in the economic stimulus plan (ARRA) promised $7.2 billion to increase the reach and use of broadband, reviving both the phrase "digital divide" and the government programs intended to remedy it.

5. Foundations, government agencies, and nonprofit groups have recently begun to use the phrase "broadband divide." See, for example, Masai 2009, which describes a National Telecommunications and Information Administration/Rural Utilities Service six-day series of public hearings and roundtables on how the federal government should spend $7.2 billion in economic stimulus funding to increase access to broadband. A full panel of the hearings focused on the "broadband have-nots." As part of this digital divide revival, legislators have deployed the language of "unserved" or "underserved" populations, "have-nots," and "digital gaps." See also Feinberg 2009.

6. The focus on distributional strategies makes sense when it is considered in political-economic context. Under neoliberalism, the state's commitment to protect the general public from the risks and vulnerabilities of the free market is displaced onto citizens as an individual and family responsibility. For example, the Personal Responsibility and Work Opportunity Reconciliation Act of 1996 reframed social welfare in terms of individual empowerment and personal responsibility in order to justify state withdrawal from providing welfare and other forms of support for poor families. Devolution, a key technique of neoliberalism, transfers federal powers to state and local governments. For example, since the mid-1990s, the federal government has increasingly relied on community development block grants—which provide a huge degree of local flexibility and discretion in spending—to fund a wide variety of antipoverty, affordable housing, and family support programs. Distributional strategies to address high-tech equity dilemmas have remained popular in part because these practices assign responsibility to individuals for navigating an increasingly tempestuous and volatile global economy when the state withdraws structural support for its citizens.

7. Articulating these insights was often a collaborative process, building on conversations that began in WYSMS meetings, workshops, and other public events and drawing on a mix of techniques. Embedded as I was in a long-term commitment to the YWCA community, my research methods emerged from specific situations, recent events, or ongoing social justice projects. Though I covered a standard list of questions—available in the methodological appendix—in more formal research settings, such as the one-on-one interviews, I often had to be responsive and develop techniques on the fly.

8. Iris Marion Young (1990) lists these as the "five faces of oppression."

9. This is a crucial point when organizing with women struggling to meet their basic needs. It is difficult to make visible just how much the state and even some anti-poverty organizations are biased against poor and working-class women within consensus models of organizing, because models that prioritize consensus have historically sought ways to contain difference and limit conflict. A politics of dissensus and difference, however, should enable the expression of difference in the interest of justice and promote increased participation, inclusion, and meaningful deliberation.

Chapter 4

1. Manufacturing sectors, particularly, have stagnated or declined since the mid-1950s. In 2000, only 7 percent of Troy's residents engaged in manufacturing activity, compared with a national average of 14 percent. Services are the largest source of employment, employing about 46 percent of the population (U.S. Census 2000). The residents' per capita income is the region's lowest—nearly two-thirds of the population are low or moderate income, and 35 percent of households in Troy make less than $20,000 a year.

2. Troy's population has declined steadily since the turn of the last century. In 1900, the population of Troy was 60,651, making it the sixty-second largest city in the United States. The population of Troy climbed as high as 81,104 by 1925. After the 1950 census, however, Troy's population began to decline; it was 72,763, in 1950, 62,007 in 1970, and 54,269 in 1990, a loss of nearly 5,000 residents per decade. In 2000, the population dipped below 50,000 for the first time in the twentieth century (to a mere 49,107, only 60 percent of its record high in 1925). It is projected to decrease still further.

3. In the early 2000s, we began to see the first appreciable reverse in the decades old trend of white flight from cities and metropolitan areas across the United States. Reverse white flight is particularly pronounced in Manhattan, 150 miles south of Troy, where the African American and Hispanic populations have stagnated or declined and median family incomes and house values have skyrocketed since 2000. Many individuals and families have been priced out of the New York City metropolitan area, and more were unsettled by the events of September 11, 2001. These two trends combined to cause a particularly notable change in Hudson Valley demographics after 2000.

4. Because Troy's nineteenth-century architecture is strikingly lovely, downtown brownstones could be bought for relatively modest prices, and the city government was eager to auction or sell off vacant buildings, the urge to rehab and flip property was strong in Troy. This was the case for individuals and families, but also for commercial developers, one of whom described himself to the local press as feeling like "a kid in a candy store" before snapping up six massive downtown buildings, three of which were major historic landmarks, in 2003 (O'Brien 2004).

5. Carole Turbin (1994) argues that these unions were able to succeed in part because the twin industries of the city, iron manufacturing and shirt-collar and cuff stitching and laundering, required a gender-differentiated and skilled workforce, which meant that men could work when the women in their families were on strike, and vice versa. Union organization in Troy was truly a family affair.

6. For more on the Charles Nalle story, see Christianson (2010). There is even a persistent but probably apocryphal folk story in Troy that when John Brown's body was being moved by train after his execution in Virginia (for attacking the federal armory at Harpers Ferry in an attempt to incite slaves to open rebellion) to his final resting place in North Elba, New York, his wife tried to have his body laid out in state in each major city they passed. The only city that actually granted permission for a public viewing of his body, my neighbors will tell you, was Troy. According to this story, his body was visited by thousands of people as it lay in state in the lobby of the American House.

7. New York's "Tech Valley" encompasses eighteen counties running along the Hudson River from Montreal to just north of New York City. It includes Albany, Clinton, Columbia, Dutchess, Essex, Franklin, Fulton, Greene, Hamilton, Montgomery, Orange, Rensselaer, Saratoga, Schenectady, Schoharie, Ulster, Warren, and Washington counties (Tech Valley Chamber Coalition 2003).

8. Reliable investment numbers are difficult to find, but regional business groups such as the Tech Valley Chamber Coalition and the Center for Economic Growth estimate that combined private and public investment in Tech Valley between 2001 and 2009 was between $10.5 and $17.6 billion. According to the Tech Valley Chamber Coalition, however, much of this was *proposed* spending, and it is difficult to ascertain how much of this investment was actually delivered in the Capital Region.

9. The 2003 summit was titled "Building a Technopolis," the 2004 event was titled "Bringing the World to Tech Valley." The event was canceled in 2008.

10. One developer, Sandy Horowitz, bought six large building in downtown Troy in a single year, including three of the city's major landmarks (O'Brien 2004). Only two years after he was proclaimed the savior of Troy by the local media, the city shut off his water for failure to pay his bills, and, by June 2006 he owed the city $249,000 in back taxes (Hardin 2006). As of this writing, in May 2010, all of his downtown buildings are in foreclosure.

11. This pattern seems to be repeating in the current recession. Though little data are available on the most recent changes, anecdotal experience suggests that home foreclosures and the Obama administration's Home Buyer's Credit has set off a new round of housing speculation in the area, while municipal and state cuts to social service programs and educational institutions have left vulnerable citizens increasingly without support.

12. There is a rich and complex academic literature that traces the impacts of volatility on workers, both domestically and internationally, in a nuanced and empirically-grounded way. See, for example, Benner 2002; Castells 1996; Castells and Aoyama 1994; Drennan 2002; Fingleton 1999; Harvey 1990; and McCall 2001.

13. "The canonical New Economy discourse . . . goes something like this. Finally, after a long wait, the computer revolution is paying off economically. . . . All that hardware, now linked from local area networks to the global Internet, along with a political regime of smaller government and lighter regulation, has unleashed forces of innovation and wealth creation like the world has never known before. Flatter hierarchies and more interesting work are the social payoffs; rising incomes and an end to slumps the economic payoffs. Quality replaces quantity, knowledge replaces physical capital, and flexible networks replace rigid organization charts" (Henwood 2003, 3–4).

14. Among the most popular are Daniel Bell's *The Coming of Post-Industrial Society* (1976), Alvin Toffler's *The Third Wave* (1980), John Naisbett's *Megatrends* (1984), Peter Drucker's *Post-Capitalist Society* (1993), Bill Gates's *The Road Ahead* (1996), Kevin Kelly's *Out of Control* (1995) and *New Rules for the New Economy* (1998), Esther Dyson's *Release 2.0* (1997), Thomas Friedman's *The World Is Flat* (2005), and Don Tapscott's *Wikinomics* (2006).

15. Volatility is increasingly the topic of policy discussions about inequality and development. As Joshua Aizenman and Brian Pinto argue in *Managing Economic Volatility and Crises: A Practitioner's Guide* (2005), there is a significant relationship between economic volatility and inequality. They write, "Cross-country studies have consistently found that volatility exerts a significant negative impact on long-run (trend) growth, which is exacerbated in poorer countries" (3). The slowed economic growth and increasing inequality that follow economic volatility, they argue, is magnified by weak political institutions, leading to permanent setbacks for developing countries in relationship to richer countries. This global relationship is replicated in the Third World/First World relationship between the poor and working class and the professional-middle and owning classes in the United States.

16. Employment insecurity, for example, is highly mediated by both structural dislocation and persistent racial and gender divisions of labor. New "configurations of inequality" attendant on the information economy are widely variable, and geographically defined. Leslie McCall (2001) explores four regions: industrial Detroit, postindustrial Dallas, high-tech St. Louis, and immigrant Miami. In industrial Detroit, class inequality among men and racial inequality among men and women are lower, while class inequality among women, gender inequality, and education-ally based wage inequalities are higher. In high-tech St. Louis, class inequality among men and women is lower while racial inequality among men and women, gender inequality, and educationally based wage inequalities are higher (McCall 2001, 49). Rather than offering a single path to lowering inequality of all kinds, the informa-

tion economy offers new, more complex forms of inequality. We must then look closely at the equity conflicts embedded in alternate paths of economic development for any given region.

17. See Silicon Valley Toxics Coalition 1997. See also Benner 2002; Pellow and Park 2002; and Pitti 2003.

18. See also Sassen 1991.

19. One possible standout in this generally grim picture is in education and health services, where both the number of jobs available and the wages for these jobs are growing steadily. The strong unions enjoyed by teachers and health care workers in New York State likely influence this trend.

20. See also Barker and Feiner 2009 and Peterson 2003.

21. In her meticulous study of inequality in the new economy, *Complex Inequality*, Leslie McCall argues that there is "clear empirical evidence of greater gender and racial inequality in high-technology regions, especially high-technology manufacturing areas. While reindustrialization efforts involving technological transformations are to be applauded for delivering a new sense of hope to devastated communities, these efforts must simultaneously be construed as investments in gender and racial inequality" (McCall 2001, 190).

22. This kind of self-discipline is a common technique of "devolutionism" in high-tech work, which uses information technologies to push responsibility downward and flatten the hierarchy of control in a company (Sewell and Wilkinson 1992; Zuboff 1989). But, as Sewell and Wilkinson write, this does not necessarily translate into more freedom or control for workers. "'Responsibility' is devolved only under the condition of a strict monitoring of compliance with instructions" (Sewell and Wilkinson 1992, 278–279). Sewell and Wilkinson argue that management regimes like these create an "electronic panopticon" that controls IT workers through technologically mediated self-surveillance. But control in the high-tech, low-wage workplace is often more straightforward, vertical, and centralized.

23. As Benner (2002) argues, the growth of flexible employment relationships in the information economy has served to undermine employers' investments in training. He writes,

Most employers are reluctant to make the investment in retraining their workforce for fear that the workers will leave and take the skills with them, or even if they stay, that the demand for skills will not remain long enough to make the investment worthwhile. The lack of retraining exacerbates imbalances between supply and demand when markets shift because employers have to wait for employees to train themselves in the new skills, typically waiting for the next generation of graduates (Cappelli 2000). Thus, rapidly changing work requirements can create significant labor market shocks for people, increasing the risk of serious misfortune. (222–223).

24. Nevertheless, the policy presumption that low-income people are information or technology poor has shown remarkable persistence. Claims that the poor "lack

knowledge, professionalism, networking . . . and training" (Attali 2000) or that low-income people lack the education to be relevant in a "skills- and knowledge-based economy" are prevalent in public policy, as well as the popular and academic press (Mossberger, Tolbert, and Stansbury 2003; Norris 2001; Servon 2002). At the turn of the twenty-first century, government agencies expressed fears that the United States risked losing the scientific, economic, and human resource advantages if the technological workforce was not significantly strengthened (Information Technology Association of America 1998; National Science Foundation 2000; President's Information Technology Advisory Committee [PITAC] 1999). Several of these agencies, including the Information Technology Association of America and the President's Information Technology Advisory Committee, argued that, in particular, "the under-representation of women and minorities in the IT workforce [was] a serious national problem" (PITAC 1999). In the face of this looming skills crisis, government agencies and nongovernmental organizations suggested strategies for "bridging the gap" through worker retraining and reskilling programs.

Chapter 5

1. See, for example, Gilliom 2001; Piven and Cloward 1971; and Tice 1998.

2. In his insightful collected volume, *Surveillance and Security: Technological Politics and Power in Everyday Life* (2006), Torin Monahan suggests that many scholars are asking the wrong questions about electronic surveillance because they fail to recognize that technologies operate not only as tools but also as creators of social worlds.

3. Winner writes that technology is legislation, and that "technical forms do, to a large extent, shape the basic pattern and content of human activity in our time." Therefore, "politics becomes (among other things) an active encounter with the specific forms and processes contained in technology." He concludes that "modern technics, much more than politics as conventionally understood, now legislates the conditions of human existence (Winner 1977, 323).

4. Lessig describes the Internet as "an exploding space of social control" where "control is coded, by commerce, with the backing of the government" (Lessig 1999, ix–x).

5. See, for example, Campbell 2005; Mettler 2005; Mettler and Soss 2004; Schneider and Ingram 1997; and Soss 1999. Policy feedback argues that the design and operation of public policy and state institutions—everything from the GI Bill to Temporary Aid to Needy Families (TANF), the local social service office to the county prison—have impacts on mass political opinion, mobilization, and behavior. Political scientists interested in policy feedback ask questions about how public policies and their implementation "render citizens more or less engaged in politics and . . . shape citizens' beliefs, preferences, demands, and power" (Mettler and Soss 2004, 5).

6. Citizenship is, then, a relationship learned in context, differentially available to women according to their social location. A different form of client-citizenship is,

for example, available to women receiving TANF than to women receiving SSDI. As Soss points out, being an AFDC (now TANF) recipient "reduces the odds that a person will vote to slightly less than half of what it would have been otherwise" (Soss 1999, 364), even if other demographic characteristics are held constant. In confirmation of Barbara Nelson's (1984) arguments about the effects of the two-tiered benefit system, Soss notes that SSDI recipients, on the other hand, are just as politically active as the rest of the citizenry. For Soss, this is a clear indication that there is a unique relationship between participation in means-tested social service programs and participation in other formal mechanisms of governance.

7. Soss's respondents perceived themselves as quite politically astute. They echoed Barbara Nelson's (1984) arguments that AFDC participation is a particularly challenging and demanding relationship with the government, and questioned whether Soss himself would be able to understand its complicated demands and irrational rules well enough to complete his article.

8. After 1996, changes to welfare policy, including the five-year lifetime benefits limit and block grant structure of TANF, require that information about clients (1) be retained over the clients' entire lifetime, (2) be shared between state systems in order to police eligibility for benefits, and (3) be shared among federal, state, and local governing bodies, nonprofit community organizations, and for-profit welfare-to-work agencies.

9. In *Medical Research for Hire: The Political Economy of Pharmaceutical Clinical Trials* (2009), Jill A. Fisher makes a similar point about the relationship between the pharmaceutical clinical trials industry and its "volunteer" research subjects. The industry responds in part to the decreasing availability of medical insurance for many Americans, and clinical trials are often the only kind of care available to the working poor in the United States, who make too much to qualify for Medicaid but too little to pay for private insurance when they are not covered by an employer. This is often the case for women, whose employment is more commonly contingent and temporary, and therefore less likely to qualify them for health insurance. The nature and extent of the research subject's ability to "consent" in this situation are questionable.

10. The only clients who have more information protection are victims of domestic violence. Victim information in public records and databases is protected under the Violence Against Women Act..

11. Freire 1973, 1997, 1998.

Chapter 6

1. I am aware that there is a great deal of concern about the appropriation or co-optation of the word "empowerment" in both scholarly and activist circles (for perhaps the best example, see Cruikshank 1999). It is certainly true that the term is often cynically and paternalistically deployed. However, I have chosen to use the

word here in the spirit in which it is used at the YWCA of Troy-Cohoes. The YWCA continues to use the word, as I do, because of its explicit recognition of power relationships. In fact, when they write empowerment, they capitalize the "power": emPOWERment. For the YWCA, emPOWERment is not something bestowed on one party by another. Rather, it is the process of exercising personal and social power to acquire the tools you need to craft the life you want for yourself.

2. In 2009, the national YWCA revised the mission to read: "The YWCA is dedicated to eliminating racism, empowering women and promoting peace, justice, freedom and dignity for all." Before 2009, however, the YWCA of Troy-Cohoes and the national organization used the mission I quote here in printed materials, staff training, and programming.

3. "Sanctioned ignorance" is a phrase coined by Gayatri Chakravorty Spivak (1999, x) to describe culturally approved holes in knowledge that are authorized and promoted to support a status quo of inequality and cultural imperialism. These are the things we are taught to ignore absolutely by mainstream education (ibid., 2). Recent feminist work on "epistemologies of ignorance," the practices that account for *not* knowing, takes this original insight in many exciting new directions, including how the "practices of ignorance are often intertwined with practices of oppression and exclusion" (Tuana and Sullivan 2006, vii). For more on epistemologies of ignorance, see *Hypatia* 21, no. 3 (Summer 2006) or *Frontiers: A Journal of Women's Studies* 30, no. 1 (2009).

4. The concept of popular education certainly did not originate with radical Brazilian educator Paulo Freire, but he is generally credited with systematizing and describing its principles most thoroughly and perceptively. His narratives of the radical democratic promise of educational practice are certainly compelling and provocative (for more on Freire's central position in popular education, see Mayo 1999 and McLaren and Leonard 1993). Though the Freirian method of popular education is the best documented and most well known in academic circles, I find the American examples of popular education uniquely interesting and woefully underappreciated.

For example, Jane Addams and Ellen Starr Gates opened Hull House in Chicago in 1889, espousing a belief that embeddedness in the vast knowledge and experience of ordinary people is critical to promoting justice and democracy and arguing that democratic citizenship relies on the development of a social rather than individual system of ethics. "The Settlement," Addams wrote, "is an experimental effort to aid in the solution of the social and industrial problems which are engendered by the modern conditions of life in a great city. It insists that these problems are not confined to any one portion of a city. It is an attempt to relieve, at the same time, the overaccumulation at one end and the destitution at the other" (Addams 1911, 98). The solution, for Addams, lay neither in professionalized social work, which she loathed, nor in charity, which she called "a deception based on impossible virtues" (28), but in true democracy. Settlements, she believed,

could build democracy by providing community spaces for collective investigation into reality, and by providing resources for grassroots reform and social
movement.

An often unacknowledged foremother of American sociology, one of Addams's
favorite clubs at Hull House was the Working People's Social Science Club. Organized
by an English workingman in 1890, the club drew 40–100 participants weekly for
seven years by allowing a speaker one hour to talk, and then opening the floor for
an hour's discussion. Addams wrote proudly and at length about it in *Twenty Years
at Hull House* as a most enthusiastic club, where "zest for discussion was unceasing,
and . . . all discussion save that which 'went to the root of things' was impatiently
discarded as an unworthy, halfway measure" (Addams 1960, 119).

Myles Horton, one of the founders of the Highlander Folk School, was directly
influenced by a 1930 visit to Hull House. He visited the settlement house several
times, "solicited Addams' advice on starting a 'Southern Mountain School'" (Smith
2003, 5), and continued to correspond with Addams in the early 1930s. Many years
later, he wrote that Addams's ideas about democracy and decision-making profoundly influenced his own. He wrote,

Jane Addams . . . once was asked for her definition of democracy. She said, "It means people
have the right to make decisions. If there is a group of people sitting around a country store
and there's a problem they're talking about, there are two ways to do it. They can go out to
some official to tell them what to do, or they can talk it out and discuss it themselves. Democracy is if they did it themselves." (Horton, Kohl, and Kohl 1998, 49)

The connection of education to the broader goal of democratic social ethics and
political change inspired Horton to begin, with Don West, the Highlander Folk
School, a residential adult education program in East Tennessee. Horton declared
that its purpose was to "train community leaders for participation in a democratic
society" and to help spread democratic principles to all human relationships in every
political, economic, social, and cultural activity.

Their programs over the last seventy-nine years—including workshops on labor
organizing and education (1930s and 1940s), desegregation in the public schools
(1950s), citizenship and voter registration (1950s and 1960s), civil rights organizing
and leadership (1960s), strip mining and toxic waste dumping in Appalachia (1970s),
economic globalization issues (1980s), and multilingual organizing, interracial
coalition building, and youth leadership in the American South (1990s to the
present)—have been incredibly ambitious. Nevertheless, Highlander is dedicated to
building decision-making processes that are based in broad participation, transparency, and accountability. I would argue that this focus on democratic process and
the recognition of the innate value of all people's experience and native intelligence
has kept Highlander ahead of the curve of every major American social movement
since its inception in the 1930s.

I owe many of my insights on Jane Addams's influence on ideas about social
democracy to Sanford Schram's wonderful book, *Praxis for the Poor: Piven and Cloward
and the Future of Social Science in Social Welfare* (2002).

5. The tradition of which Addams, Horton, and Freire are a part has been arguing for one hundred years that advocacy, on its own, is incompatible with creating a just democratic order. Addams couched her argument about the obsolescence of epistemological charity in a moral imperative: society loses when any individual is not educated so that her moral power is made available to the world. Horton was more pragmatic, writing that "advocacy alone, while titillating and intellectually stimulating, has never brought about radical change. It has an important role to play only when coupled with a dynamic process of transferring decision-making powers to the population most immediately concerned" (Horton 1972, 28). Freire writes that the point of departure for a movement toward liberation must be the people themselves, who can begin to see their situations as challenging, rather than merely limiting, through collective inquiry. However, he writes, "If people, as historical beings necessarily engaged with other people in a movement of inquiry, did not control that movement, it would be (and is) a violation of their humanity. Any situation in which some individuals prevent others from engaging in the process of inquiry is one of violence. The means used are not important; to alienate human beings from their own decision-making is to change them into objects" (Freire 1970, 66).

6. See, for example, Bjerknes and Bratteteig 1995; Gustavsen 2001; and Levinger 1998.

7. See the Black Panther Party's March 1972 platform, or the "Ten Point Program," point 6. See also Alondra Nelson's forthcoming book, *Body and Soul: The Black Panther Party and the Politics of Health and Race* (Berkeley: University of California Press). On participatory research in the women's health movement, see, for example, Morgen 2002. On AIDS movement activism, see, for example, Epstein 1998.

8. See, for example, Fals-Borda and Rahman 1991; Fischer 2000; and Rahman 1993.

9. See also Ehn 1992.

10. See, for example, Balka 1997a, 1997b, 2000; Balka and Smith 2000; Greenbaum and Kyng 1991; Gregory 2000; Rothschild 1998, 1999; and Suchman 1987.

11. The project was similar to Randall Pinkett's "Community Connections" project in Roxbury, Massachusetts (<http://communityconections.org>) and the "Neighborhood Knowledge Los Angeles" and "I Am LA" projects of the UCLA Advanced Policy Institute (<http://nkla.ucla.edu>).

12. Paul Kivel offers an insightful critique of the "social service" model of community organizing in his essay, "Social Service or Social Change?" He argues that social service work in the movement to end violence against women addresses the needs of individuals reeling from the personal and devastating impact of institutional systems of exploitation and violence, while social change work challenges the root causes of the exploitation and violence. While social change work is oriented toward

creating social movement and fomenting structural change, he argues, social service work provides a professionalized "buffer zone" that controls individuals needing services, contains their protest, and co-opts community leadership (Kivel 2007).

13. I say that these are the most challenging parts of the Resource Directory project because functionality like "Talk Back" would give social service clients the ability to evaluate and critique community programs and organizations publicly, and to share information among themselves. This would no doubt raise issues and unearth information about the operation of programs that some social service organizations might find threatening.

14. The calculator is available online at <http://www.wceca.org>.

15. Papadopoulos, Scanlon, and Lees used reconstructed stories as a method of presenting and validating findings from interview data obtained during the Enfield Vision Research Project. Researchers in this project conducted ninety interviews with visually impaired people. After the first twenty interviews were completed, researchers presented findings to participants in the form of reconstructed stories developed according to themes and subthemes (including "registering or not registering as blind or partially sighted," "social activities," and "mobility and transport") that emerged from data analysis. Respondents were asked to reflect on these stories on the basis of several questions:

• What are your general impressions of the stories?
• Do you identify with the issues in the stories?
• Are the stories true and credible to you?
• Are any of the issues in the stories strange to you?
• Do you disagree with any of the stories?
• Are any stories missing?

Research participants were then able to respond to the research findings and confirm their credibility and dependability (Papadopoulos, Scanlon, and Lees 2002, 277).

16. Interview questions included:

• What is"the system"?
• What keeps people from getting on [the social service system] when they need to?
• What keeps people from getting off [the social service system] when they want to?
• How could it be easier to get some of the services?
• What skills do you need to navigate the system?
• What patterns (with income limits, etc.) have you experienced?
• How much do you think you know about social services on a scale of 1–10?
• Are there things that you would like to know more about what social services provides?

17. For more on policy parables, see the Southern Rural Development Institute's "Parables to Policy" project at <http://www.srdi.org>.

18. For example, both welfare reform and the digital divide mark a historical turn away from structural redistributive solutions to those based on the role of individual citizens as consumers (of government services such as public assistance or of informational products and devices). Policymaking in both areas takes place in a political and social context marked by neoliberalism, devolution, and decentralization, which profoundly influences both how problems are understood and how solutions are imagined and articulated.

19. Freire 1998; Giroux 2001.

Chapter 7

1. Notable exceptions include Barney and Gordon 2005; Frankenfeld 1992; Franklin 1999; Laird 1993 Leach, Scoones, and Wynne 2005; Lessig 1999; Sclove 1995; Winner 1977, 1979, 1986; and Zimmerman 1995.

2. Berger makes a similar point about the mostly African American, HIV-positive activists she worked with in Detroit while writing her lovely book, *Workable Sisterhood* (2004).

3. I have included some sample agendas in appendix B.

4. For more on the tension between friendship and equality in engaged feminist work, see Francesca Polletta's wonderful book, *Freedom Is an Endless Meeting* (2002), especially her discussion of the women's movement's attempts to balance the personal and the political through the practice of "sisterhood" on pages 149–175.

5. He writes, "The 'social work' of everyday mutual aid is a critical dimension of the politics of state building and therefore how a 'politics of survival' that grows out of everyday concerns is related to a 'politics of social change' that focuses on long-run structural transformation" (Schram 2002, 35).

6. In "Situated Knowledges: The Science Question in Feminism and the Privilege of Partial Perspective," a chapter in her pathbreaking work, *Simians, Cyborgs and Women*, Donna Haraway writes, "I am arguing for politics and epistemologies of location, positioning, and situating, where partiality and not universality is the condition of being heard to make rational knowledge claims. These are claims on people's lives . . . only the god-trick [of seeing everything from nowhere] is forbidden" (Haraway 1991, 195).

7. For more on strong objectivity and socially situated knowledge, see Sandra Harding's work, including *The Feminist Standpoint Theory Reader: Intellectual and Political Controversies* (2004), particularly chapter 8, "Rethinking Standpoint Epistemology: What Is 'Strong Objectivity'?," and *Is Science Multicultural? Postcolonialisms, Femi-*

nisms, and Epistemologies (1998). She writes, strong objectivity "provide[s] a kind of method for maximizing our ability to block 'might makes right' in the sciences. Maximizing objectivity is not identical to maximizing neutrality . . . in a certain range of cases, maximizing neutrality is an obstacle to maximizing objectivity" (129).

8. See also Daniel Ray White's discussion of ecosociality and engaged objectivity in *Postmodern Ecology: Communication, Evolution and Play* (White 1998, 178–180).

9. Shiv Visvanathan makes a similar argument about the role of science and grass-roots people's science movements in India (Visvanathan 2005, 91).

10. This insight comes primarily from the work of Donna Haraway. For Haraway, there simply are no innocent powers with which, or positions from which, to represent the world. Rather, "objectivity turns out to be about particular and specific embodiment . . . only partial perspective promises objective vision" (Haraway 1991, 190). For Haraway, objectivity is always "a local achievement . . . [concerned with] holding things together well enough so that people can share in that account powerfully" (Haraway and Goodeve 2000, 161).

Conclusion

1. For more on consensus conferences, see, for example, Anderson and Jaeger 2002 and Grundahl 1995.

2. For more information on science shops, see the Web site of Living Knowledge: The International Network of Science Shops at <http://www.scienceshops.org>. Also of interest are Gnaiger and Martin 2001; van de Vusse 1985; and Wallace 2001.

3. Historical information about LINC can be found at <http://www.lincproject.org>.

4. The Partnership for Working Families has a wonderful online resource for more information about community benefits agreements at <http://www.community benefits.org>.

5. ARISE's Equity Agenda is available at <http://www.ariseorg.net/about_us.html>.

6. For more on ARISE's work, see <http://www.ariseorg.net>.

7. Silicon Valley has twenty-nine Superfund sites, the densest concentration in the United States. In their splendid book, *The Silicon Valley of Dreams* (2002), David Naguib Pellow and Lisa Sun-Hee Park argue that between 1950 and 2001, the Santa Clara Valley committed "ecocide." The region aggressively courted the electronics sector to replace the waning canning industry and agricultural economy that had long dominated the valley, and the policy of San Jose business and municipal leaders was growth at any cost. By the early 1980s, the dirty secrets of the pristine-seeming, landscaped technology manufacturing "campuses" and "parks" were leaking out— literally. In 1981, it was discovered that a drinking well that supplied 16,500 homes

in San Jose was contaminated with trichloroethane (TCA), a toxic solvent used in the production of microchips. TCA had been leaking from a faulty underground storage tank that stored nearly 60,000 gallons of various toxic waste materials produced by semiconductor production. In the following years, more spills were discovered, including one near an IBM semiconductor plant that remains one of the largest toxic spills ever to occur in the United States.

Appendix A

1. A significant portion of the Highlander Archives is held at the Wisconsin State Historical Archives and the Tennessee State Library and Archives.

2. I write at length about this aspect of the research in "Double-Bound: Putting the Power Back in Participatory Research" (Eubanks 2009).

3. Feminist scholars have had a vigorous ongoing discussion about negotiating ethical and empirical commitments in research with human subjects. Particularly interesting is Gesa E. Kirsch's "Friendship, Friendliness, and Feminist Fieldwork," which is worth quoting at length on the issue of consent in participatory research projects:

I believe that we need to develop more realistic—and perhaps more limited—expectations about relationships with participants in both service-learning and research projects. . . .We may want to consider introducing such important concepts as "confirming consent," a notion proposed by Paul V. Anderson (1998, 75), who suggests that when participants find themselves in particularly vulnerable positions . . . they ought to be given the opportunity to renegotiate consent after the fieldwork is completed. . . .We may also want to introduce the "right to co-interpretation" (Newkirk 1996, 13), a concept advanced by Thomas Newkirk, who proposes that we should offer our emerging interpretations of research data to participants for their review and comments. (Kirsch 2005, 2168)

References

Abraham, Kaye. 1994. *Mama Might Be Better Off Dead: The Failure of Health Care in Urban America*. Chicago: University of Chicago Press.

Abramowitz, Mimi. 1996. *Regulating the Lives of Women: Social Welfare Policy from Colonial Times to the Present*. Boston: South End Press.

Abramowitz, Mimi. 2000. *Under Attack, Fighting Back: Women and Welfare in the United States*. New York: Monthly Review Press.

Addams, Jane. [c. 1910] 1960. *Twenty Years at Hull House*. New York: Signet.

Addams, Jane. 1911. *Democracy and Social Ethics*. Urbana: University of Illinois Press.

Aizenman, Joshua, and Brian Pinto. 2005. *Managing Economic Volatility and Crises: A Practitioner's Guide*. Oxford: Cambridge University Press.

Anderson, Ida-Elizabeth, and Birgit Jaeger. 2002. Danish Participatory Models: Scenario Workshops and Consensus Conferences: Towards More Democratic Decision-Making. *Pantaneto Forum* 6. <http://www.pantaneto.co.uk/issue6/andersenjaeger.htm> (accessed May 17, 2010).

Arms, John, Albert Roundtree, and Buck Maggard. 1969. Ten Commandments for Outside Organizers. Group I, Box I, Series II, File 6 (General Highlander Materials). New Market, TN: Highlander Research and Education Center Archives.

Arnstein, Sherry R. 1969. A Ladder of Citizen Participation. *Journal of the American Planning Association* 35:216–224.

Attali, J. 2000. A Market Solution to Poverty: Microfinance and the Internet. *New Perspectives Quarterly* 17:31–33.

Balka, Ellen. 1997a. *Computer Networking: Spinsters on the Web*. Ottawa: Canadian Research Institute for the Advancement of Women.

Balka, Ellen. 1997b. Participatory Design in Women's Organizations: The Social World of Organizational Structure and the Gendered Nature of Expertise. *Gender, Work and Organization* 4:99–115.

Balka, Ellen. 1999. Where Have All the Feminist Technology Critics Gone? *Loka Alert* 6.

Balka, Ellen, and Richard Smith. 2000. *Women, Work and Computerization: Charting a Course to the Future*. New York: Kluwer.

Banta, Martha. 1995. *Taylored Lives: Narrative Productions in the Age of Taylor, Veblen, and Ford*. Chicago: University of Chicago Press.

Barker, Drucilla, and Susan F. Feiner. 2004. *Liberating Economics: Feminist Perspectives on Families, Work, and Globalization*. Ann Arbor: University of Michigan Press.

Barker, Drucilla, and Susan F. Feiner. 2009. Affect, Race, and Class: An Interpretive Reading of Caring Labor. *Frontiers: A Journal of Women's Studies* 30:41–54.

Barney, Darin. 2000. *Prometheus Wired: The Hope for Democracy in the Age of Network Technology*. Chicago: University of Chicago Press.

Barney, Darin, and Aaron Gordon. 2005. Education and Citizenship in the Digital Age. *Techne: Research in Philosophy and Technology* 9.

Bastian, Michelle. 2006. Haraway's Lost Cyborg and the Possibilities of Transversalism. *Signs: Journal of Women in Culture and Society* 31:1027–1049.

Bastian, Sunil, and Nicola Bastian. 1996. *Assessing Participation: A Debate from South Asia*. New Delhi: Konark Publishers.

Bell, Daniel. 1976. *The Coming of Post-Industrial Society*. New York: Basic Books.

Benhabib, Seyla. 1996. *Democracy and Difference: Contesting the Boundaries of the Political*. Princeton, NJ: Princeton University Press.

Benner, Chris. 2002. *Work in the New Economy: Flexible Labor Markets in Silicon Valley*. Oxford: Blackwell Publishers.

Berger, Michele Tracy. 2004. *Workable Sisterhood: The Political Journey of Stigmatized Women with HIV/AIDS*. Princeton, NJ: Princeton University Press.

Bernhardt, Annette, and Christine Owens. 2009. Rebuilding a Good Jobs Economy. *The Nation*, March 30. Available online at <http://www.thenation.com/doc/20090330/bernhardt_owens> (accessed June 5, 2009).

Besser, Howard. 2001. The Next Digital Divides. *Teaching to Change LA* 1.

Bjerknes, Gro, and Tone Bratteteig. 1995. User Participation and Democracy. A Discussion of Scandinavian Research on System Development. *Scandinavian Journal of Information Systems* 7:73–98.

Blum, Lenore, and Carol Frieze. 2005. The Evolving Culture of Computing: Similarity Is the Difference. *Frontiers: A Journal of Women Studies* 26:110–125.

Braman, Sandra. 2007. *Change of State: Information, Policy, and Power*. Cambridge, MA: MIT Press.

Brown, Mark Malloch. 2001. Can ICTs Address the Needs of the Poor? *Choices* 10:4.

Buckberg, E., and A. Thomas. 1995. Wage Dispersion and Job Growth in the United States. *Finance & Development* (June):16–19.

Bullard, Robert D. 2000. *Dumping in Dixie: Race, Class, and Environmental Quality*. 3rd ed. Boulder, CO: Westview Press.

Campbell, Andrea Louise. 2005. *How Policies Make Citizens: Senior Political Activism and the American Welfare State*. Princeton, NJ: Princeton University Press.

Campbell, Nancy D. 2000. *Using Women: Gender, Drug Policy and Social Justice*. New York: Routledge.

Campbell, Nancy D., and Virginia Eubanks. 2004. Making Sense of Imbrication: Popular Technology and "Inside-Out" Methodologies. In *Proceedings of the Participatory Design Conference 2004*, 65–73. Toronto, ON and New York: Computer Professionals for Social Responsibility and ACM.

Castells, Manuel. 1996. *The Rise of the Network Society: The Information Age: Economy, Society, and Culture*. Malden, MA: Blackwell.

Castells, Manuel, and Yuko Aoyama. 1994. Paths Towards the Informational Society: Employment Structures in G-7 Countries, 1920–1990. *International Labour Review* 133:5–33.

Children's Partnership. 2002. *Online Content for Low-income and Underserved Americans: An Issue Brief by the Children's Partnership*. Santa Monica, CA: Children's Partnership.

Christianson, Scott. 2010. *Freeing Charles: The Struggle to Free a Slave on the Eve of the Civil War*. Urbana: University of Illinois Press.

Citron, Danielle Keats. 2008. Technological Due Process. *Washington University Law Review* 85.

City of Troy. 2005. 5 Year Consolidated Plan, 2005–2009. Troy, NY. <http://www.troyny.gov/projects/consolidatedplan.html> (accessed December 21, 2009).

Clement, Anne, and Leslie R. Shade. 2000. The Access Rainbow: Conceptualizing Universal Access to the Information/Communication Infrastructure. In *Community Informatics: Enabling Communities with Information and Communication Technologies*, ed. Michael Gurstein, 32–51. Hershey, PA: Idea Group Publishing.

Cohoon, J. McGrath, and William Aspray. 2008. *Women and Information Technology: Research on Underrepresentation*. Cambridge, MA: MIT Press.

Collingwood, Harris. 2003. The Sink-Or-Swim Economy. *New York Times Magazine,* June 8, 42–45.

Collins, H. M., and Robert Evans. 2002. The Third Wave of Science Studies: Studies of Expertise and Experience. *Social Studies of Science* 32:235–296.

Collins, Patricia Hill. 1986. Learning from the Outsider Within: The Sociological Significance of Black Feminist Thought. *Social Problems* 33:14–32.

Collins, Patricia Hill. 1989. The Social Construction of Black Feminist Thought. *Signs: Journal of Women in Culture and Society* 14:745–773.

Collins, Patricia Hill. 1990. *Black Feminist Thought: Knowledge, Consciousness, and the Politics of Empowerment.* New York: Routledge.

Collins, Patricia Hill. 1998. *Fighting Words: Black Women and the Search for Justice.* Minneapolis: University of Minnesota Press.

Compaine, Benjamin M., ed. 2001. *The Digital Divide: Facing a Crisis or Creating a Myth?* Cambridge, MA: MIT Press.

Cooke, Bill, and Uma Kothari. 2001. *Participation: The New Tyranny?* London: Zed Books.

Cooks, Leda, and Kristen Isgro. 2005. The "Cyber Summit" and Women: Incorporating Gender into Information and Communication Technology UN Policies. *Frontiers: A Journal of Women Studies* 26:71–89.

Cozzens, Susan. 2007. Distributive Justice in Science and Technology Policies. *Science & Public Policy* 34:85–94.

Crenshaw, Kimberlé. 1991. Mapping the Margins: Intersectionality, Identity Politics, and Violence against Women of Color. *Stanford Law Review* 43:1241–1299.

Cruikshank, Barbara. 1999. *The Will to Empower: Democratic Citizens and Other Subjects.* Ithaca: Cornell University Press.

DataCenter. 2007. Research Justice Initiative. Oakland, CA. <http://www.datacenter.org/programs/ResearchJustice.pdf> (accessed June 3, 2009).

DeBoer, Larry, and Michael C. Seeborg. 1989. The Unemployment Rates of Men and Women: A Transition Probability Analysis. *ILR Review* 42 (3):404–414.

Denner, Jill, Linda Werner, Steve Bean, and Shannon Campe. 2005. The Girls Creating Games Program: Strategies for Engaging Middle-School Girls in Information Technology. *Frontiers: A Journal of Women Studies* 26:90–98.

Dickard, Norris. 2002. *Federal Retrenchment on the Digital Divide: Potential National Impact.* Washington, DC: Benton Foundation.

Domestic Workers United. 2009. Domestic Worker's Bill of Rights. <http://www .domesticworkersunited.org/campaigns.php> (accessed May 17, 2010).

Drennan, Matthew. 2002. *The Information Economy and American Cities*. Baltimore, MD: Johns Hopkins University Press.

Drucker, Peter F. 1993. *Post-Capitalist Society*. New York: HarperCollins.

Dyson, Esther. 1997. *Release 2.0*. New York: Broadway.

Ehn, Pelle. 1992. Scandinavian Design: On Participation and Skill. In *Usability: Turning Technologies into Tools*, ed. P. S. Adler and T. A. Winograd, 96–132. New York: Oxford University Press.

Epstein, Stephen. 1998. *Impure Science: AIDS, Activism, and the Politics of Knowledge*. Berkeley: University of California Press.

Eubanks, Virginia. 2006. Technologies of Citizenship: Surveillance and Political Learning in the Welfare System. In *Surveillance and Security: Technology and Power in Everyday Life*, ed. Torin Monahan. New York: Routledge.

Eubanks, Virginia. 2007. Trapped in the Digital Divide: The Distributive Paradigm in Community Informatics. *The Journal of Community Informatics* 3.

Eubanks, Virginia. 2009. Double-Bound: Putting the Power Back in Participatory Research. *Frontiers: A Journal of Women's Studies* 30 (1):107–137.

Fals Borda, Orlando, and Muhammad Anisur Rahman. 1991. *Action and Knowledge: Breaking the Monopoly with Participatory Action-Research*. New York: Apex Press.

Farkas, Nicole. 2002. Bread, Cheese and Expertise: Dutch Science Shops and Democratic Institutions. PhD diss., Science and Technology Studies, Rensselaer Polytechnic Institute, Troy, New York.

Farmer, Paul. 1999. *Infections and Inequalities: The Modern Plagues*. Berkeley: University of California Press.

Farmer, Paul. 2005. *Pathologies of Power: Health, Human Rights, and the New War on the Poor*. Berkeley: University of California Press.

Feenberg, Andrew. 1990. The Ambivalence of Technology. *Sociological Perspectives* 33:35–50.

Feenberg, Andrew. 1991. *Critical Theory of Technology*. New York: Oxford University Press.

Feinberg, Andrew. 2009. Legislators See "Underserved" Definition as First Step for Broadband Stimulus. <http://broadbandbreakfast.com/2009/04/legislators-see -underserved-definition-as-first-step-for-broadband-stimulus> (accessed May 17, 2010).

Fingleton, Eamonn. 1999. *In Praise of Hard Industries: Why Manufacturing, Not the Information Economy, Is the Key to Future Prosperity.* Boston: Houghton Mifflin.

Fischer, Frank. 2000. *Citizens, Experts, and the Environment: The Politics of Local Knowledge.* Durham, NC: Duke University Press.

Fisher, Jill A. 2009. *Medical Research for Hire: The Political Economy of Pharmaceutical Clinical Trials.* New Brunswick, NJ: Rutgers University Press.

Flyvbjerg, Bent. 2001. *Making Social Science Matter: Why Social Inquiry Fails and How It Can Succeed Again.* New York: Cambridge University Press.

Folbre, Nancy. 2001. *The Invisible Heart: Economics and Family Values.* New York: New Press.

Frankenfeld, Phillip J. 1992. Technological Citizenship: A Normative Framework for Risk Studies. *Science, Technology & Human Values* 17:459–484.

Franklin, Ursula. 1999. *The Real World of Technology.* Toronto, ON: House of Anansi.

Fraser, Nancy. 1989. *Unruly Practices: Power, Discourse and Gender in Contemporary Social Theory.* Minneapolis: University of Minnesota Press.

Fraser, Nancy. 1997. *Justice Interruptus: Critical Reflections on the 'Postsocialist' Condition.* New York: Routledge.

Freire, Paulo. 1973. *Education for Critical Consciousness.* New York: Continuum.

Freire, Paulo. [1970] 1997. *Pedagogy of the Oppressed.* New York: Continuum.

Freire, Paulo. 1998. *Pedagogy of Freedom: Ethics, Democracy, and Civic Courage.* New York: Rowman & Littlefield.

Friedman, Thomas L. 2005. *The World Is Flat: A Brief History of the Twenty-First Century.* New York: Picador.

Fuller, Laurie, and Erica Meiners. 2005. Reflections: Empowering Women, Technology, and (Feminist) Institutional Changes. *Frontiers: A Journal of Women Studies* 26:168–180.

Fusco, Coco. 2002. *The Bodies That Were Not Ours and Other Writings.* New York: Routledge.

Gans, H. J. 1995. Fitting the Poor into the New Economy. *Technology Review* 98:72–73.

Gates, Bill. 1996. *The Road Ahead: Completely Revised and Up-to-Date.* New York: Penguin Books.

Gibbs, N. R. 1995. Working Harder, Getting Nowhere. *Time*, July 3, 16–20.

Gilliom, John. 2001. *Overseers of the Poor: Surveillance, Resistance, and the Limits of Privacy*. Chicago: University of Chicago Press.

Giroux, Henry A. 2001. *Theory and Resistance in Education: Towards a Pedagogy for the Opposition*, rev. ed.. Westport, CT: Bergin & Garvey Paperback.

Glaser, Barney G. 1998. *Doing Grounded Theory: Issues and Discussions*. Mill Valley, CA: Sociology Press.

Glaser, Barney G., and Anselm L. Strauss. 1967. *The Discovery of Grounded Theory: Strategies for Qualitative Research*. Chicago: Aldine.

Gnaiger, A., and E. Martin. 2001. Science Shops: Operational Options: Report #1 of the EU Research Project SCIPAS—Study and Conference on Improving Public Access to Science through Science Shops. Ultrecht, Netherlands: Science Shop for Biology, Ultrecht University.

Gonzalez, Susan. 2001. Director Spike Lee Slams "Same Old" Black Stereotypes in Today's Films. *Yale Bulletin and Calendar* 29.

Goodwin, Jeff, James M. Jasper, and Francesca Polletta, eds. 2001. *Passionate Politics: Emotions and Social Movements*. Chicago: University of Chicago Press.

Gore, Al. 1995. The GII: Conditions for Success. *Intermedia* 23:48.

Goslee, Susan. 1998. *Losing Ground Bit by Bit: Low-Income Communities in the Information Age*. Washington, DC: Benton Foundation.

Greenbaum, Joan. 1993. A Design of One's Own: Towards Participatory Design in the United States. In *Participatory Design: Principles and Practices*, ed. Douglas Schuler and Aki Namioka. Hillsdale, NJ: Lawrence Erlbaum Associates.

Greenbaum, Joan, and Morten Kyng. 1991. *Design at Work*. Hillsdale, NJ: Lawrence Erlbaum Associates.

Gregory, Judith. 2000. Sorcerer's Apprentice: Inventing the Electronic Health Record, Reinventing Patient Care. Department of Communication, University of California, San Diego.

Grundahl, Johs. 1995. The Danish Consensus Conference Model. In *Public Participation in Science: The Role of Consensus Conferences in Europe*, ed. Simon Joss and John Durant. London: NMSI Trading, Ltd.

Gurstein, Michael. 2003. Effective Use: A Community Informatics Strategy Beyond the Digital Divide. *First Monday* 8 (12). <http://firstmonday.org/htbin/cgiwrap/bin/ojs/index.php/fm/article/view/1107/1027> (accessed May 17, 2010).

Gustavsen, B. 2001. Theory and Practice: The Mediating Discourse. In *Handbook of Action Research: Participative Inquiry & Practice*, ed. Peter Reason and Hilary Bradbury, 17–26. Thousand Oaks, CA: Sage.

Haraway, Donna. 1991. *Simians, Cyborgs and Women: The Reinvention of Nature*. New York: Routledge.

Haraway, Donna. 1997. *Modest Witness@Second Millenium. FemaleMan© Meets Onco-Mouse™: Feminism and Technoscience*. New York: Routledge.

Haraway, Donna, and Thyrza Nichols Goodeve. 1999. *How Like a Leaf: An Interview with Donna J. Haraway*. New York: Routledge.

Hardin, Chet. 2004. Can It Happen Here? *Metroland* 29 (6). <http://www.metroland.net/back_issues/vol29_no26/features.html>.

Harding, Sandra. 1998. *Is Science Multicultural? Postcolonialisms, Feminisms, and Epistemologies*. Bloomington: Indiana University Press.

Harding, Sandra, ed. 2004. *The Feminist Standpoint Theory Reader: Intellectual and Political Controversies*. New York: Routledge.

Harding, Sandra. 2006. *Science and Social Inequality: Feminist and Postcolonial Issues*. Chicago: University of Illinois Press.

Harding, Sandra, and Karen Norberg. 2005. New Feminist Approaches to Social Science Methodologies: An Introduction. *Signs: Journal of Women in Culture and Society* 30:2009–2015.

Harvey, David. 1990. *The Condition of Postmodernity: An Enquiry into the Origins of Cultural Change*. Malden, MA: Blackwell.

Harvey, David. 2008. The Right to the City. *New Left Review* 53:23–40.

Hawkesworth, Mary. 1988. *Theoretical Issues in Policy Analysis*. Albany: State University Press of New York.

Henwood, Doug. 2003. *After the New Economy: The Binge and the Hangover That Won't Go Away*. New York: New Press.

Hobson, Janell. 2007. Searching for Janet in Cyberspace: Race, Gender, and the Interface of Technology. In *Techknowledgies: New Imaginaries in Humanities, Arts, and Techno-sciences*, ed. Mary Valentis. London: Cambridge Scholars Press.

hooks, bell. 1994. *Teaching to Transgress: Education as the Practice of Freedom*. New York: Routledge.

Hopper, Kim. 2003. *Reckoning with Homelessness*. Ithaca, NY: Cornell University Press.

Horton, Myles. 1933. Educational Theory: Mutual Education. Group 2, Series I, Box 2, File 39 (Miles Horton Papers). New Market, TN: Highlander Research and Education Center Archives.

Horton, Myles. 1954. Transcript of Tape Recording on Leadership Prepared by the Workshop Committee. Group 2, Series I, Box 7, File 117 (Miles Horton Papers). New Market, TN: Highlander Research and Education Center Archives.

Horton, Myles. c. 1972. Decision-Making Processes. Group 2, Series I, Box 8, File 137 (Miles Horton Papers). New Market, TN: Highlander Research and Education Center Archives.

Horton, Myles, and Paulo Freire. 1990. *We Make the Road by Walking: Conversations on Education and Social Change*. Philadelphia: Temple University Press.

Horton, Myles, Judith Kohl, and Herbert Kohl. 1998. *The Long Haul: An Autobiography*. New York: Teachers College Press.

Hossfeld, Karen J. 2001. "Their Logic Against Them": Contradictions in Sex, Race and Class in Silicon Valley. In *TechniColor: Race, Technology, and Everyday Life*, ed. Alondra Nelson, Thuy Linh N. Tu, and Alicia Hines, 34–63. New York: New York University Press.

Information Technology Association of America. 1998. Building the 21st Century Information Technology Workforce: Underrepresented Groups in the Information Technology Workforce. Task Force Report, Information Technology Association of America.

Jessup, Elizabeth, and Tamara Sumner. 2005. Design-Based Learning and the Participation of Women in IT. *Frontiers: A Journal of Women Studies* 26:141–147.

Jobs with Justice. 2009. Unions Matter. <http://www.jwj.org/freechoice/union smatter/index.html> (accessed May 17, 2010).

Kabir, Naila, ed. 2005. *Inclusive Citizenship: Meanings and Expressions*. London: Zed Books.

Karr, Timothy. 2008. Obama's Broadband Roadmap. *Huffington Post*. <http://www .huffingtonpost.com/timothy-karr/obamas-broadband-roadmap_b_149321.html> (accessed Nov. 3, 2009).

Kekelis, Linda S., Rebecca Wepsic Ancheta, and Etta Heber. 2005. Hurdles in the Pipeline: Girls and Technology Careers. *Frontiers: A Journal of Women Studies* 26:99–109.

Kelly, Kevin. 1995. *Out of Control: The New Biology of Machines, Social Systems, & the Economic World*. New York: Basic Books.

Kelly, Kevin. 1998. *New Rules for the New Economy*. New York: Penguin.

Kesby, Mike. 2005. Retheorizing Empowerment-through-Participation as a Performance in Space: Beyond Tyranny to Transformation. *Signs: Journal of Women in Culture and Society* 30:2037–2065.

Killen, Andreas. 2006. *Berlin Electropolis: Shock, Nerves, and German Modernity*. Berkeley: University of California Press.

Kirsch, Gesa E. 2005. Friendship, Friendliness, and Feminist Fieldwork. *Signs: Journal of Women in Culture and Society* 30:2163–2172.

Kivel, Paul. 2007. Social Service or Social Change? In *The Revolution Will Not Be Funded: Beyond the Non-Profit Industrial Complex*, ed. INCITE! Women of Color Against Violence, 129–149. Cambridge, MA: South End Press.

Kolko, Beth E. 2000. *Race in Cyberspace*. New York: Routledge.

Komar, Brian. 2003. Civil Rights Coalition Applauds Preservation of Funding for Federal Community Technology Programs. <http://civilrights.org> (accessed June 2, 2010).

Kropotkin, Petr Alekseevich. 1902. *Mutual Aid: A Factor of Evolution*. New York: McClure Phillips.

Kumar, Amitav. 2001. Temporary Access: The Indian H-1B Worker in the United States. In *TechniColor: Race, Technology, and Everyday Life*, ed. Alondra Nelson, Thuy Linh N. Tu, and Alicia Hines, 76–87. New York: New York University Press.

Labaton, Stephen. 2001. New F.C.C. Chief Would Curb Agency Reach. New York Times Online, February 7, 2001. <http://www.nytimes.com/2001/02/07/technology/07FCC.html?printpage=yes> (accessed Dec 19, 2009).

Laird, Frank N. 1993. Participatory Analysis, Democracy, and Technological Decision Making. *Science, Technology & Human Values* 18:341–361.

Lather, Patti. 2007. *Getting Lost: Feminist Efforts toward a Double(d) Science*. Albany: State University of New York Press.

Lather, Patti, and Chris Smithies. 1997. *Troubling the Angels: Women Living with HIV/ AIDS*. Boulder, CO: Westview Press.

Leach, Melissa, Ian Scoones, and Brian Wynne, eds. 2005. *Science and Citizens: Globalization and the Challenge of Engagement*. London: Zed Books.

Ledwith, Margaret. 2001. Community Work as Critical Pedagogy: Re-envisioning Freire and Gramsci. *Community Development Journal* 36:171–182.

Lessig, Lawrence. 1999. *Code: And Other Laws of Cyberspace*. New York: Basic Books.

Levinger, David. 1998. Participatory Design History. <http://cpsr.org/prevsite/conferences/pdc98/history.html> (accessed May 17, 2010).

Lorde, Audre. 1984. *Sister Outsider*. New York: Crossing Press.

Lugones, María, and Joshua Price. 1995. Certainty, Simplicity, and Agreement: The Cognitive Basis of Monoculturalism. In *Dominant Culture: El Deseo por un Alma Pobre*, ed. D. A. Harris, 103–127. Westport, CT: Bergin & Garvey.

Lynch, Michael, and Simon Cole. 2005. Science and Technology Studies on Trial: Dilemmas of Expertise. *Social Studies of Science* 35:269–311.

Mack, Raneta Lawson. 2001. *The Digital Divide: Standing at the Intersection of Race and Technology*. Durham, NC: Carolina Academic Press.

Margolis, Jane. 2008. *Stuck in the Shallow End: Education, Race, and Computing*. Cambridge, MA: MIT Press.

Margolis, Jane, and Allan Fisher. 2002. *Unlocking the Clubhouse: Women in Computing*. Cambridge, MA: MIT Press.

Marx, Leo. 1993. Does Improved Technology Mean Progress? In *Technology and the Future*, 6th ed., ed. Albert H. Teich, 3–14. New York: St. Martin's Press.

Masai, Jesse. 2009. Definitions and Broadband Measures Must Evolve. Account for On-the-Ground Realities. <http://broadbandbreakfast.com/2009/03/definitions-and-broadband-measures-must-evolve-account-for-on-the-ground-realities> (accessed May 17, 2010).

Mayer-Schönberger, Viktor, and David Laxer. 2007. *Governance and Information Technology: From Electronic Government to Information Government*. Cambridge, MA: MIT Press.

Mayo, Peter. 1999. *Gramsci, Freire and Adult Education: Possibilities for Transformative Action*. New York: Palgrave Macmillan.

McCall, Leslie. 2001. *Complex Inequality: Gender, Class, and Race in the New Economy*. New York: Routledge.

McLaren, Peter, and Peter Leonard. 1993. *Paulo Freire: A Critical Encounter*. New York: Routledge.

Mettler, Suzanne. 2005. *Soldiers to Citizens: The G.I. Bill and the Making of the Greatest Generation*. New York: Oxford University Press.

Mettler, Suzanne, and Joe Soss. 2004. The Consequences of Public Policy for Democratic Citizenship: Bridging Policy Studies and Mass Politics. *Perspectives on Politics* 2:1–19.

Mink, Gwendolyn. 1998. *Welfare's End*. Ithaca, NY: Cornell University Press.

Monahan, Torin. 2006. *Surveillance and Security: Technological Politics and Power in Everyday Life*. New York: Routledge.

Morgen, Sandra. 1983. The Politics of "Feeling": Beyond the Dialectic of Thought and Action. *Women's Studies* 10:203–223.

Morgen, Sandra. 1995. "It Was the Best of Times, It Was the Worst of Times": Emotional Discourse in the Work Cultures of Feminist Health Clinics. In *Feminist Organizations: Harvest of the New Women's Movement,* ed. Marx Ferree and Patricia Yancey Martin. Philadelphia: Temple University Press.

Morgen, Sandra. 2002. *Into Our Own Hands: The Women's Health Movement in the United States, 1969–1990.* New Brunswick, NJ: Rutgers University Press.

Mossberger, Karen, Caroline J. Tolbert, and Mary Stansbury. 2003. *Virtual Inequality: Beyond the Digital Divide.* Washington, DC: Georgetown University Press.

Naisbett, John. 1984. *Megatrends.* Spennymoor, Durham, UK: Macdonald Press.

Nakamura, Lisa. 2002. *Cybertypes: Race, Ethnicity, and Identity on the Internet.* New York: Routledge.

Nakamura, Lisa. 2007. *Digitizing Race: Visual Cultures of the Internet.* Minneapolis: University of Minnesota Press.

Naples, Nancy. 2003. *Feminism and Method: Ethnography, Discourse Analysis, and Activist Research.* New York: Routledge.

National Institute for Literacy. 2002. Senate Proposes to Restore Funding For Even Start, Prison Literacy, and Job Training Programs. <http://www.nifl.gov/texis/search/context.html?query=Isserlis&pr=global&prox=page&rorder=500&rprox=500&rdfreq=500&rwfreq=500&rlead=500&rdepth=0&sufs=0&order=r&cq=&cmd=context&id=4615968a1> (accessed May 17, 2010).

National Science Foundation (NSF). 2000. Dear Colleague Letter. Information Technology Workforce. Dear Colleague Letter, NSF 00-77. Washington, DC: National Science Foundation.

National Telecommunications and Information Administration. 1995. *Falling Through the Net: A Survey of the "Have Nots" in Urban and Rural America.* Washington, DC: U.S. Department of Commerce.

National Telecommunications and Information Administration. 1998. *Falling Through the Net II: New Data on the Digital Divide.* Washington, DC: U.S. Department of Commerce.

National Telecommunications and Information Administration. 1999. *Falling Through the Net: Defining the Digital Divide.* Washington, DC: U.S. Department of Commerce.

National Telecommunications and Information Administration. 2000. *Falling Through the Net: Toward Digital Inclusion.* Washington, DC: U.S. Department of Commerce.

National Telecommunications and Information Administration. 2002. *A Nation Online: How Americans are Expanding Their Use of the Internet*. Washington, DC: U.S. Department of Commerce.

Nelson, Alondra. Forthcoming. *Body and Soul: The Black Panther Party and the Politics of Health and Race*. Berkeley: University of California Press.

Nelson, Alondra, Thuy Linh N. Tu, and Alicia Hines. 2001. *TechniColor: Race, Technology and Everyday Life*. New York: New York University Press.

Nelson, Barbara. 1984. Women's Poverty and Women's Citizenship: The Political Consequences of Economic Marginality. *Signs: Journal of Women in Culture and Society* 10:209–231.

Newell, Peter, and Joanna Wheeler, eds. 2006. *Rights, Resources and the Politics of Accountability*. London: Zed Books.

Ng, Cecilia, and Swasti Mitter. 2005. *Gender and the Digital Economy: Perspectives from the Developing World*. Thousand Oaks, CA: Sage Publications.

Norris, Pippa. 2001. *Digital Divide: Civic Engagement, Information Poverty, and the Internet Worldwide*. New York: Cambridge University Press.

Novak, Thomas P., and Donna L. Hoffman. 1998. Bridging the Digital Divide: The Impact of Race on Computer Access and Internet Use. Working paper. <http://wwww2000.ogsm.vanderbilt.edu/papers/race/science.html>.

Oates, Stephen B. 1994. *Let the Trumpet Sound: A Life of Martin Luther King, Jr*. New York: HarperPerennial.

Obama, Barack H. 2009. Address to the Congress. Remarks as Prepared for Delivery. <http://www.guardian.co.uk/world/2009/feb/25/full-text-barack-obama-congress-address> (accessed May 17, 2010).

Obey, Dave, and Committee on Appropriations. 2009. Summary: American Recovery and Reinvestment. <http://appropriations.house.gov/pdf/PressSummary01-15-09.pdf> (accessed May 17, 2010).

O'Brien, Tim. 2004. Reinventing Troy. *Rensselaer* Spring.

O'Connor, Alice. 2001. *Poverty Knowledge: Social Science, Social Policy, and the Poor in Twentieth-Century U.S. History*. Princeton, NJ: Princeton University Press.

OMBWatch. 2002. Community Technology Programs Cut Back in FY '03 Budget. <http://www.ombwatch.org/article/articleview/220/1/78> (accessed May 17, 2010).

Papadopoulos, Irene, Karen Scanlon, and Shelley Lees. 2002. Reporting and Validating Findings through Reconstructed Stories. *Disability & Society* 17: 269–281.

Pellow, David Naguib, and Lisa Sun-Hee Park. 2002. *The Silicon Valley of Dreams: Environmental Injustice, Immigrant Workers, and the High-Tech Global Economy.* New York: New York University Press.

Perera, Gihan. 2008. Claiming Rights to the City. *Who Owns Our Cities?* 15(1).

Perkins, Douglas D., and Abraham Wandersman. 1990. "You'll Have to Work to Overcome Our Suspicions": The Benefits and Pitfalls of Research with Community Organizations. *Social Policy* 21:32–41.

Peterson, V. Spike. 2003. *A Critical Rewriting of Global Political Economy: Integrating Reproductive, Productive and Virtual Economies.* New York: Routledge.

Pew Research Center. 2003. Ever-Shifting Internet Population: A New Look at Internet Access and the Digital Divide. A Pew Internet and American Life Project report. Washington DC: Pew Research Center. <http://www.pewinternet.org/Reports/2003/The-EverShifting-Internet-Population-A-new-look-at-Internet-access-and-the-digital-divide.aspx> (accessed May 17, 2010).

Pitti, Stephen J. 2003. *The Devil in Silicon Valley: Northern California, Race, and Mexican Americans.* Princeton, NJ: Princeton University Press.

Piven, Frances Fox, and Richard Cloward. 1971. *Regulating the Poor: The Functions of Public Welfare.* New York: Vintage.

Polletta, Francesca. 2002. *Freedom Is an Endless Meeting: Democracy in American Social Movements.* Chicago: University of Chicago Press.

President's Information Technology Advisory Committee. 1999. Information Technology Research: Investing in Our Future. Report to the President. <http://www.ccic.gov/ac/report> (accessed July 20, 2004).

Puckett, Jim, Leslie Byster, Sarah Westervelt, Richard Gutierrez, Sheila Davis, Asma Hussain, and Madhumitta Dutta. 2002. Exporting Harm: The High-Tech Trashing of Asia. Basel Action Network and Silicon Valley Toxics Coalition. <http://www.ban.org/E-waste/technotrashfinalcomp.pdf> (accessed May 17, 2010).

Rahman, Muhammad Anisur. 1993. *People's Self-Development: Perspectives on Participatory Action Research: A Journey Through Experience.* London: Zed Books.

Reardon, Kenneth M. 2000. An Experiential Approach to Creating Community/University Partnership That Works: The East St. Louis Action Research Project. *Cityscape: A Journal of Policy Development and Research* 5:59–74.

Rischard, Jean-Francois. 1996. Connecting Developing Countries to the Information Technology Revolution. *SAIS Review* (Winter–Spring):93–107.

Roberts, Dorothy. 1998. *Killing the Black Body: Race, Reproduction, and the Meaning of Liberty.* New York: Vintage.

Roberts, Dorothy. 2003. *Shattered Bonds: The Color of Child Welfare*. New York: Basic Civitas Books.

Ross, Loretta J. 2005. A Feminist Perspective on Katrina. *ZNet: The Spirit of Resistance Lives*. <http://www.zcommunications.org/a-feminist-perspective-on-katrina-by-loretta-j-ross> (accessed May 17, 2010).

Rosser, Sue Vilhauer. 2005. Through the Lens of Feminist Theory: Focus on Women and Information Technology. *Frontiers: A Journal of Women Studies* 26:1–23.

Rothschild, Joan. 1998. Designed Environments and Women's Studies: A Wake-Up Call. *NWSA Journal* 10:17.

Rothschild, Joan. 1999. *Design and Feminism: Re-Visioning Spaces, Places and Everyday Things*. New Brunswick, NJ: Rutgers University Press.

Sandoval, Chela. 1995. New Sciences: Cyborg Feminism and the Methodology of the Oppressed. In *The Cyborg Handbook*, ed. Chris Hables Gray. New York: Routledge.

Sassen, Saskia. 1991. *The Global City: New York London Tokyo*. Princeton, NJ: Princeton University Press.

Schneider, Anne L., and Helen M. Ingram. 1993. Social Construction of Target Populations: Implications for Politics and Policy. *American Political Science Review* 87:334–347.

Schneider, Anne L., and Helen M. Ingram. 1997. *Policy Design for Democracy*. Lawrence: University Press of Kansas.

Schram, Sanford. 2002. *Praxis for the Poor: Piven and Cloward and the Future of Social Science in Social Welfare*. New York: New York University Press.

Schuler, Douglas, and Aki Namioka. 1993. *Participatory Design: Principles and Practices*. Hillsdale, NJ: Lawrence Erlbaum Associates.

Sclove, Richard. 1995. *Democracy and Technology*. New York: Guilford Press.

Scott-Dixon, Krista. 2004. *Doing IT: Women Working in Information Technology*. Toronto, ON: Sumach Press.

Service Employees International Union. 2009. Our Union. <http://www.seiu.org/a/ourunion/fast-facts.php> (accessed May 29, 2009).

Servon, Lisa. 2002. *Bridging the Digital Divide: Technology, Community, and Public Policy*. Malden, MA: Blackwell Publishing.

Sewell, Graham. 1998. The Discipline of Teams: The Control of Team-Based Industrial Work Through Electronic and Peer Surveillance. *Administrative Science Quarterly* 43:397–428.

Sewell, Graham, and Barry Wilkinson. 1992. "Someone to Watch Over Me": Surveillance, Discipline, and the Just-in-Time Labor Process. *Sociology: The Journal of the British Sociological Association* 26:271–289.

Silicon Valley Toxics Coalition. 1997. Silicon Valley Toxic Tour. <http://www.svtc.org/site/PageServer?pagename=svtc_silicon_valley_toxic_tour> (accessed May 17, 2010).

Silliman, Jael, Marlene Gerber Fried, Loretta Ross, and Elena Gutierrez. 2004. *Undivided Rights: Women of Color Organize for Reproductive Justice.* Cambridge, MA: South End Press.

Smith, Angela. 2003. Myles Horton, Highlander Folk School and the Wilder Coal Strike of 1932. http://mtsu.academia.edu/AngelaSmith/Papers/87740/Myles-Horton--Highlander-Folk-School--and-the-Wilder-Coal-Strike-of-1932 (accessed May 17, 2010).

Smith, Dorothy. 1974. Women's Perspective as a Radical Critique of Sociology. *Sociological Inquiry* 44:7–13.

Smith, Dorothy. 1987. *The Everyday World as Problematic.* Toronto, ON: University of Toronto Press.

Smith, Dorothy. 1999. *Writing the Social: Critique, Theory, and Investigations.* Toronto, ON: University of Toronto Press.

Smith, Dorothy. 2005. *Institutional Ethnography: A Sociology for People.* New York: Rowman Altamira.

Soss, Joe. 1999. Lessons of Welfare: Policy Design, Political Learning, and Political Action. *American Political Science Review* 93:363–380.

South End Press Collective. 2007. *What Lies Beneath: Katrina, Race, and the State of the Nation.* Cambridge, MA: South End Press.

Spivak, Gayatri Chakravorty. 1999. *A Critique of Postcolonial Reason: Toward a History of the Vanishing Present.* Cambridge, MA: Harvard University Press.

Standing, Guy. 1999. Global Feminization Through Flexible Labor: A Theme Revisited. *World Development* 27:583–602.

Stephenson, John B. 2008. Electronic Waste: EPA Needs to Better Control Harmful U.S. Exports through Stronger Enforcement and More Comprehensive Regulation. U.S. Government Accountability Office, Washington, DC. <http://www.gao.gov/new.items/d081044.pdf> (accessed May 17, 2010).

Stoeker, Randy. 2005. Is Community Informatics Good for Communities? Questions Confronting an Emerging Field. *Journal of Community Informatics* 1.

Stone, Deborah. 1988. *Policy Paradox: The Art of Political Decision Making.* New York: HarperCollins.

Subramaniam, Banu. 2009. Moored Metamorphoses: A Retrospective Essay on Feminist Science Studies. *Signs: Journal of Women in Culture and Society* 34:951–980.

Suchman, Lucy A. 1987. *Plans and Situated Actions: The Problem of Human-Machine Communications*. Cambridge: Cambridge University Press.

Tapscott, Don. 2006. *Wikinomics: How Mass Collaboration Changes Everything*. New York: Portfolio Hardcover.

Tarjanne, P. 1995. The GII: Moving Towards Implementation. *Telecommunications* 29.

Tech Valley Chamber Coalition. 2003. Welcome to New York's Tech Valley. <http://www.techvalley.org/regioninfo.php?PHPSESSID=c175a488e21573e405e17e8bdb3de31f> (accessed July 22, 2004).

Thompson, Becky. 2001. *A Promise and a Way of Life: White Antiracist Activism*. Minneapolis: University of Minnesota Press.

Tice, Karen. 1998. *Tales of Wayward Girls and Immoral Women: Case Records and the Professionalization of Social Work*. Urbana: University of Illinois Press.

Tillberg, Heather K., and J. McGrath Cohoon. 2005. Attracting Women to the CS Major. *Frontiers: A Journal of Women Studies* 26:126–140.

Tjerandsen, Carl. 1980. *Education for Citizenship: A Foundation's Experience*. Santa Cruz, CA: Emil Schwarzhaupt Foundation.

Toffler, Alvin. 1980. *The Third Wave*. New York: Bantam.

Tuana, Nancy, and Shannon Sullivan. 2006. Introduction: Feminist Epistemologies of Ignorance. *Hypatia* 21:vii–ix.

Turbin, Carol. 1994. *Working Women of Collar City: Gender, Class, and Community in Troy, 1864–86*. Urbana: University of Illinois Press.

Turkle, Sherry. 1997. *Life on the Screen: Identity in the Age of the Internet*. New York: Simon & Schuster.

United Nations Development Programme. 1999. Globalization with a Human Face. http://hdr.undp.org/en/reports/global/hdr1999/ (accessed June 5, 2009).

U.S. Census Bureau. 1990. *Profile of selected social characteristics,* Summary File 3 (SF 3)—Sample Data. <http://factfinder.census.gov/servlet/DTGeoSearchByListServlet?ds_name=DEC_1990_STF3_&_lang=en&_ts=292168334319> (accessed December 21, 2009).

U.S. Census Bureau. 2000. *Profile of selected social characteristics,* Summary File 3 (SF 3)—Sample Data. <http://factfinder.census.gov/servlet/DTGeoSearchByListServlet?ds_name=DEC_2000_SF3_U&_lang=en&_ts=292168230952> (accessed December 21, 2009).

U.S. Census Bureau. 2008. American Community Survey 3-Year Estimates, 2006-2008. <http://factfinder.census.gov/servlet/DTGeoSearchByListServlet?ds_name=ACS_2008_3YR_G00_&_lang=en&_ts=292169330513> (accessed December 21, 2009).

United States Department of Labor. Bureau of Labor Statistics, 2008. Quarterly Census of Employment and Wages. <http://data.bls.gov:8080/PDQ/outside.jsp?survey=en> (accessed December 21, 2009).

van de Vusse, Annemarie. 1985. Visitors to the "Science Shops": Their Images of Science. Letter. *Science, Technology & Human Values* 10:75–76.

Virtual Technologies, Inc.. 2002. VirtualArrest: Offender Tracking Global Network. Press release, March 4. Spokane, WA: Virtual Technologies, Ltd.

Visvanathan, Shiv. 2005. Knowledge, Justice and Democracy. In *Science and Citizens: Globalization and the Challenge of Engagement*, ed. Melissa Leach, Ian Scoones, and Brian Wynne, 83–96. New York: Zed Books.

Wallace, Helen. 2001. The Issue of Framing and Consensus Conferences. *PLA Notes* 40:61–63.

West Harlem Environmental Action, Inc. 2002. Advancing Environmental Justice Through Community-Based Participatory Research. *EHP Supplement* 110.

White, Daniel Ray. 1998. *Postmodern Ecology: Communication, Evolution and Play*. Albany, NY: SUNY Press.

Wilhelm, Tony, Delia Carmen, and Megan Reynolds. 2002. *Connecting Kids to Technology: Challenges and Opportunities*. Baltimore, MD: Annie E. Casey Foundation.

Williams, Lee. 1997. *Grassroots Participatory Research*. Knoxville: Community Partnership Center, University of Tennessee.

Winner, Langdon. 1977. *Autonomous Technology: Technics-Out-of-Control as a Theme in Political Thought*. Cambridge, MA: MIT Press.

Winner, Langdon. 1979. Technology as Legislation. In *Technology and Change: A Courses by Newspaper Reader*, ed. J. G. Burke and M. C. Eakin, 454–461. San Francisco: Boyd and Fraser Publishing Co.

Winner, Langdon. 1986. *The Whale and the Reactor: A Search for Limits in an Age of High Technology*. Chicago: University of Chicago Press.

Wright, Michelle M. 2005. Finding a Place in Cyberspace: Black Women, Technology and Identity. *Frontiers: A Journal of Women Studies* 26:48–59.

Wyatt, Sally, Flis Henwood, Nod Miller, and Peter Senker. 2000. *Technology and In/equality: Questioning the Information Society*. New York: Routledge.

Wylie, Alison. 2003. Why Standpoint Matters. In *Science and Other Cultures: Issues in Philosophies of Science and Technology*, ed. Robert Figueroa and Sandra Harding, 26–48. New York: Routledge.

Wynne, Brian. 1996. Misunderstood Misunderstandings: Social Identities and the Public Uptake of Science. In *Misunderstanding Science? The Public Reconstruction of Science and Technology*, ed. Alan Irwin and Brian Wynne, 19–46. Cambridge: Cambridge University Press.

Young, Iris Marion. 1990. *Justice and the Politics of Difference*. Princeton, NJ: Princeton University Press.

Young, Iris Marion. 1997. Difference as a Resource for Democratic Communication. In *Deliberative Democracy: Essays on Reason and Politics*, ed. James Bohman and William Rehg, 383–406. Cambridge, MA: MIT Press.

Yunus, M. 2001. Microcredit and IT for the Poor. *New Perspectives Quarterly* 18 (1): 25–26.

Zimmerman, Andrew D. 1995. Toward a More Democratic Ethic of Technological Governance. *Science, Technology & Human Values* 20:86–107.

Zuboff, Shoshana. 1989. *In the Age of the Smart Machine: The Future of Work and Power*. New York: Basic Books.

Index

264 Index

Price, Joshua, 147
Public assistance programs, 163. *See also* Social service system
Public housing, 167
Puckett, Jim, 169

Race
  earnings, 70
  and education, 57–58
  and employment, 69–70
  and inequality, 67, 71
  and poverty status, 63
Raitano, Zianaveva, 127, 134, 136, 140–142, 156
Real-world technology, 31
Reardon, Kenneth M., 106
"Rebuilding a Good Jobs Economy," 162
Reconstructed stories, 120, 123
Redistribution, 8
Reform Organization of Welfare (ROWEL), 11
Rensselaer Polytechnic Institute, 83
Rental housing, 52
Rent control, 166–167
Reproductive justice, 29
Research, 164–165
Reynolds, Cathy, 72–73, 91
Rose, Jenn, 42, 45, 108, 139, 143–144
Ross, Loretta J., 79

Sacrificial girls, 19, 169–170
Sally Catlin Resource Center, 5
Sanctioned ignorance, 2
Sanctuary for Independent Media, 145
Scandinavia, 106
Scanlon, Karen, 120
Schneider, Anne L., 96
Schram, Sanford, 140
Schuler, Douglas, 106
Science shops, 164–165
Service Employees International Union (SEIU), 158

Service industries, 65, 75–77, 160–162
Settlement House movement, 105
Silicon Valley, 3–4, 51, 61, 168
Sims, The, 119
Single parents, 162
"Sink-or-Swim Economy, The," 53
Skill sharers, 115
Social contract, 132
Social groups, 26
Social justice, 25–26, 29
  cognitive justice, 147–148, 151–152, 163
  and distributive paradigm, 36, 45, 81
  and high-tech pollution, 168–169
  and information economy, 71, 154
  and IT, 37, 84–85, 147–148
  and popular technology, 126–127
Social location, 23–25, 27–29, 57, 148, 150
Social privilege, 42
Social reproduction, 75
Social Security Disability Insurance (SSDI), 85–86
Social service system
  double binds, 125
  and information abuse, 92–93
  and IT, 89–90
  and knowledge fragmentation, 93–94
  and limiting options, 90–91
  and narrative context, 94–95
  and participation, 97
  and political learning, 85–86, 97
  and surveillance technologies, 82–83
  target populations, 96–97
  and tracking behavior, 90
  and transparency, 91–92
  use and disclosure statement, 92–93
  and women, 30, 96–98, 120–123, 133
Soss, Joe, 85–86, 97
Standing, Guy, 71
Standpoint, 148, 150–151